D0927799

Gravity and Kinetic Energy

Developed at
The Lawrence Hall of Science,
University of California, Berkeley
Published and distributed by
Delta Education,
a member of the School Specialty Family

1465673
978-1-62571-178-6
Printing 1 – 3/2017
Webcrafters, Madison, WI

Table of Contents

Garden snails are the world's slowest land animals, averaging slightly faster than 1 cm per second. They rely on camouflage and a hard shell, not speed, for protection from predators.

How Fast Do Things Go?

Walking after a rainfall, you spot a snail. Is she perched on that rock, standing still? Or is she moving along, too slow to see? How fast is fast?

Humans use slow movements for careful work. Our daily movements, such as walking and talking, are an **average speed**. Some of our quickest movements are the blink of an eye, or a sudden sneeze. But to a hungry mosquito, we must seem slow. And to a creeping snail, we must seem fast.

The sloth holds the record for the slowest land animal. They are clumsy on the ground because of weak hind legs and long front claws. They spend most of their lives in the treetops, where these adaptations are helpful.

Slowest Animals

Animal	Estimated top speed	
	meters per minute	feet per minute
Dwarf seahorse	0.082	0.025
Banana slug	0.56	0.17
Starfish	5.2	1.6
Three-toed sloth	6.6	2.0
Tortoise	15	4.6

Life in Motion

Most animals can move. Even coral that is fixed in place most of its life floats in the ocean in its early days of life. Some animals move slowly to conserve **energy**. Some animals move fast to catch food or escape from a predator. The tables on these two pages show some information about the slowest and fastest animals in the world.

Did You Know?

Almost all countries use metric units to measure distance (meters, kilometers). The United States is one of the few places in the world to use a different system (inches, feet, miles). The global science community uses the metric system.

Fastest Animals

Animal	Estimated top speed		Record
	kilometers per hour	miles per hour	
Peregrine falcon	389	242	Fastest bird/animal
Horsefly	145	90	Fastest insect
Black marlin	129	80	Fastest animal in water
Cheetah	120	75	Fastest land animal

Did You Know?

Tuna usually cruise around 3–7 kilometers (km) per hour (2-4 mph) but can swim double that, nearly 15 km per hour (9 mph), for some time. When they are chasing prey or avoiding a hungry shark, they can accelerate faster than a sports car and reach speeds of 70 km per hour (43 mph)! It is no surprise that the word *tuna* comes from a Greek word meaning "to rush."

Not only is the cheetah the fastest land animal, this muscular big cat can accelerate from 0 to 100 km per hour in 3 seconds, beating the quickest sports cars. But one thing it can't do is climb trees!

The Fastest Human

Usain Bolt (1986–) has run faster than any other human ever recorded. The Jamaican sprinter holds the world record for fastest time in the men's 100- and 200-meter (m) sprints, and led his 4 × 100 m relay team to the world record. He is the first track athlete to complete a "triple triple," winning first place in these three events at three world championships. He won gold medals in these events at the 2008, 2012, and 2016 Olympic Games.

Bolt's Average Speed in His 2009 World Record Races

Event	Distance (meters)	Time (seconds)	Average speed over total distance (meters per second)	Average speed converted (kilometers per hour)
100 m sprint	100	9.58	10.44	37.58
200 m sprint	200	19.19	10.42	37.51

Usain Bolt, in his signature "lightning bolt" celebration pose, is the most decorated sprinter of all time. Very tall for a sprinter, Bolt takes two to four fewer strides per race than his competitors.

A stopwatch is one way to measure the time it takes a runner to travel a certain distance. In many races, finish-line electronic timing is triggered by a chip the runner wears.

Women's World Records for Running

Event	**Distance** (meters)	**Runner**	**Time** (seconds)	**Average speed over total distance** (meters/second)	**Average speed converted** (kilometers per hour)
100 m sprint	100	Florence Griffith Joyner (USA)	10.49	9.5	34.2
¼ mile	400	Marita Koch (East Germany)	47.6	8.4	30.2
½ mile	800	Jarmila Kratochvílová (Czechoslovakia)	113.3	7.1	25.6
1 K	1,000	Svetlana Masterkova (Russia)	149	6.7	24.1
1 mile	1,609	Svetlana Masterkova (Russia)	252.6	6.4	23.0
5 K	5,000	Tirunesh Dibaba (Ethiopia)	851.2	5.9	21.2
Marathon	42,195	Paula Radcliffe (United Kingdom)	8,125	5.2	18.7

Examine the table of women's records for running. What patterns do you notice in the running **speed** as the **distance** increases? Why do you think that is the case? If you graph each runner's speed, which graph would have the steepest **slope?**

Did You Know?

The fastest woman on record is Florence Griffith Joyner. No other female runner has come close to her world records, set in 1988, for the 100 m and 200 m sprints.

Speed Technology

Consider the challenge of traveling the 4,700 km from New York City, New York, to San Francisco, California. Trains of the early 1800s chugged along at 6 km per hour (4 miles per hour [mph]), just 2 km per hour (1 mph) faster than the average walking pace. In 1857, it took 4 weeks to travel across the country by train.

By the mid-1900s, passenger trains were whizzing along at 200 to 300 km per hour (124 to 186 mph). In 2015, the Japanese maglev set a world record for the fastest train, rocketing along its track at 603 km per hour (375 mph).

But nothing on the ground can compare to traveling in the sky. Modern passenger airplane speeds average about 900 km per hour (560 mph). At this speed, a trip between New York and San Francisco takes only 6 hours.

Supersonic speed is achieved when an object moves faster than the speed of sound (Mach 1). In dry air at sea level, this speed is approximately 343 meters per second (m/s), or 1,235 km per hour (767 mph). The Concorde was a supersonic passenger jet that operated between 1976 and 2003. It had a maximum speed of Mach 2.04 (2,180 km per hour, or 1,355 mph). At this speed, the Concorde could get a person from New York to San Francisco in about 2.5 hours! Although the craft could fly routes in less than half the time of other planes, they were expensive to build and maintain and are no longer in service.

The average cruising speed of today's long-range airplanes isn't as fast as it could be. The main reason is fuel economy: going faster eats up more fuel.

Speed in Space

Outside Earth's atmosphere, specially designed spacecraft can reach even higher speeds. In 1969, *Apollo 10* reached a top speed of 40,216 km per hour (nearly 25,000 mph) during its return trip from the Moon. Apollo 10 still holds the record for fastest crewed spacecraft.

To send probes farther into outer space in a reasonable amount of time, spacecraft must travel even faster. For example, even though *Juno* zipped along at an average speed of 140,013 km per hour (87,000 mph), the spacecraft took 5 years to reach Jupiter in 2016.

For an uncrewed spacecraft, *Helios 2* holds the speed record set in the mid-1970s. **Orbiting** the Sun closer than the planet Mercury, its closest approach pulled *Helios 2* into a top speed of 70.2 kilometers per second (km/s) (252,792 km per hour, or 157,078 mph).

In 2018, a NASA mission called Solar Probe Plus will be launched. It will go closer to the Sun than the Helios mission. Its orbital speed may reach 200 km/s (720,000 km per hour, or 450,000 mph). At that speed, you could travel from Earth to the Moon in less than an hour, and New York to San Francisco in less than half a minute.

Solar Probe Plus will be a historic space mission, flying into the Sun's upper atmosphere for the first time. Despite its mind-boggling speed, the probe will take 6 years to reach its target.

The Fastest Thing in the Universe

What's the fastest thing we know of? Light. The speed of light in empty space is 299,800 km/s (over 1 billion km per hour, or 670 million mph). Light provides an important measurement standard for modern physics. Because light travels at a set speed, scientists can measure distance by how far light would travel in a certain amount of time. So a distance can be listed in light-years or light-minutes.

Speed Records

	Record holder	Speed (kilometers per hour)	Speed (miles per hour)
Speed of light		1,079,280,000	670,633,500
Spacecraft (uncrewed)	*Helios 2* (1976)	252,792	157,078
Speed of Earth in orbit around the Sun		107,280	66,700
Spacecraft (crewed)	*Apollo 10* (1969)	40,216	24,989
Plane	North American X-15 rocket-powered aircraft (1967)	7,273	4,519
Speed of sound (in air)		1,236	768
Land-speed record	*Thrust SSC* car (1997)	1,228	763
Train	Japanese maglev (2015)	603	375
Recorded wind	Oklahoma tornado (1999)	511	318
Animal in air	Peregrine falcon	389	242
Human-powered vehicle	Dutch cyclist Fred Rompelberg (1995, behind a pace car)	269	167
Animal in water	Black marlin	129	80
Animal on land	Cheetah	120	75
Human on land	Usain Bolt (2009)	44	27

Think Questions

1. **Sarah the cheetah ran a 100 m dash in 5.95 s at the Cincinnati Zoo, setting a new world record for mammals. What was Sarah's speed in meters per second?**
2. **A hawk flew 600 m in 1 minute. A sparrow flew 400 m in 30 s. Which bird was faster?**
3. **If you swim the same distance in less time, are you going faster or slower? Use the speed equation and an example to explain.**

Faster and Faster

If you have ever ridden a roller coaster, you probably remember that first hill . . . the big one.

The car clatters slowly up the long slope, finally reaching the peak. After a brief pause, the track seems to go straight down. Before you know it, you are going so fast that you think you might break the sound barrier. At the foot of the first hill, you change direction and start back up the other side. Your speed changes rapidly, and you can catch your breath again.

A roller coaster is exciting because of **acceleration**. The **motion** on the roller coaster is not like the motion of a little train that chugs along at a **constant speed**. The roller coaster car changes speed. As long as the car keeps going downhill, it keeps going faster and faster. Change of speed is acceleration.

Remember the uphill run after the initial plunge? The car slowed as it went higher and higher. Again, the car changed speed, so it was once again accelerating. It may seem strange at first that speeding up and slowing down are both acceleration. But what matters to the scientist is change of speed. If it changes speed, it is accelerating. Speeding up is positive acceleration; slowing down is negative acceleration, or **deceleration**.

As these roller coaster cars climb, plunge, and roll through loops, they are constantly accelerating, or changing speed. That's what makes the ride so thrilling that we want to experience it again and again.

A bowling ball travels only slightly faster when it is released than when it collides with the pins. It rolls down the lane with very little deceleration because the lane surface is designed to reduce friction.

Where Is Acceleration in Your Life?

Just about everything in motion is accelerating. Constant speed is extremely rare. Even things that seem to travel at a steady rate really do not. Cars in a town with lots of stoplights are always speeding up and slowing down, or accelerating. Major-league fastballs slow down as they approach the batter. High fly balls slow down as they rise and speed up as they fall. Skateboarders surge forward with each push on the ground and slow between them. And when you skateboard on a hill, it is easy to accelerate out of control if you head straight down.

Can you think of anything that travels at constant speed? Toy cars can run at fairly constant speeds. Bowling balls and hockey pucks are fairly constant over moderate distances. Once a train gets going, it may continue at a constant speed until it nears its stop.

Scientists like to count and measure things, including acceleration. But when a runner sprints out of the starting blocks, what do you count or measure? To answer this, we first have to understand what acceleration is. We already know that acceleration is a change in speed. And speed depends on **change in position**. So let's start with change of position.

Many major league "power pitchers" throw fastballs clocked at 153–161 km per hour. Even if the ball decelerates by 16 km per hour between the pitcher's mound and home plate, the batter still has only about 0.35 seconds to swing the bat and make contact.

How Far?

When something is at an initial **position** and some time later it is at a final position, we say it moved. Its position is indicated by the symbol x. The Greek symbol Δ **(delta)** means change. So Δx indicates a change in position. We can measure the change of position in meters, kilometers, millimeters, or whatever units are appropriate. In a 100-meter (m) race, a runner travels from 0 m to 100 m, so Δx = 100 m.

How Fast?

Measuring how far something moved does not tell us how fast it moved. We also need to measure time. Once we know how long it took an object to change position, we can figure out speed.

Speed is a rate. Five apples for $1 is a rate. Another way to say this is five apples per dollar. Thirty miles to the gallon is a rate. Your car can travel 30 miles for every 1 gallon of gas it uses.

Earning $150 in 10 hours is not an hourly rate. To determine an hourly rate of pay, you have to figure out how much you get paid for 1 hour of work. Divide the total money into 10 equal parts: $150/10 = $15. You earn $15 for every hour you work, so $15 per hour is the hourly rate.

Scientists use the speed equation to calculate average speed when change in position and time are known. Speed is indicated by the symbol v.

$$v = \frac{\Delta x}{\Delta t}$$

A world-class sprinter like Usain Bolt can run 100 m in 10 seconds (s). If he started at $x = 0$ m and ended at $x = 100$ m, we can calculate his average speed.

$$v = \frac{\Delta x}{\Delta t} = \frac{100 \text{ m}}{10 \text{ s}} = 10 \text{ m/s}$$

Bolt ran 100 m in 10 s, so he averaged 10 m every second.

Did Bolt travel at a constant speed of 10 meters per second (m/s) all the way through the race? When the race started, he was standing still. He was going 0 m/s. That first second, he definitely moved less than 10 m. That means he must have run more than 10 m during one or more 1 s **intervals** later in the race. He did not travel at a constant speed.

When an object is traveling at a constant speed, it covers the same distance every second. Do objects that start out slow and go faster and faster also have a rate? The answer is yes. The rate at which speed changes over time is acceleration.

Usain Bolt and other 100 m sprinters typically reach their fastest speeds after running 30 m and begin to decelerate after 70 m. Bolt just slows down more slowly than anyone else!

Train Position Over Time

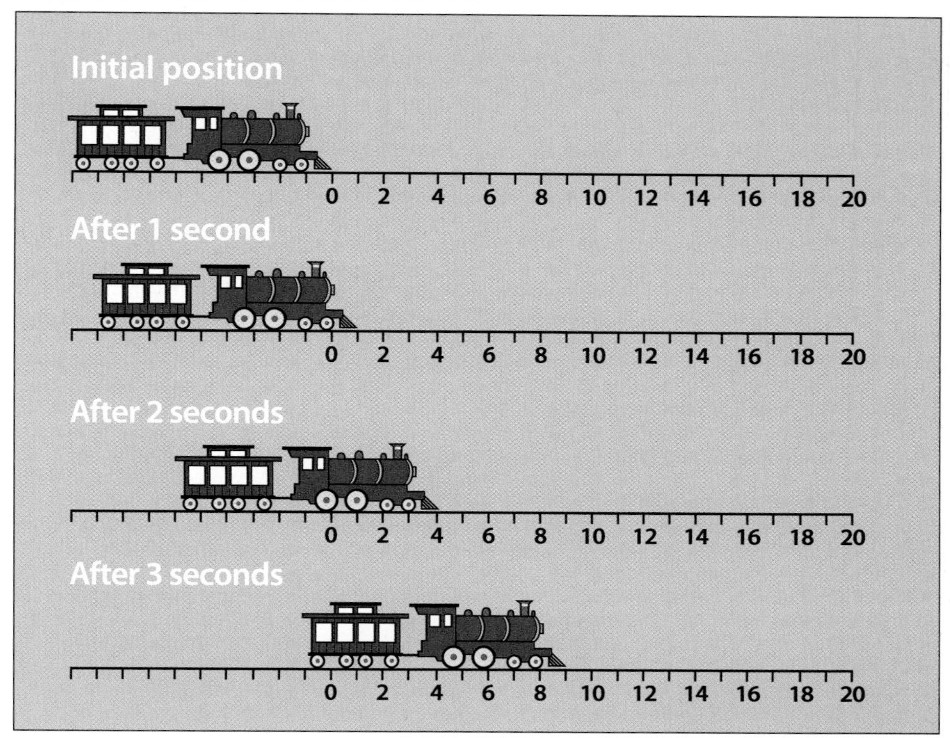

During the third second, the train traveled from the 4 m to the 9 m mark. What was the train's speed in the third second?

Changing Speed

You measure acceleration the same way you measure speed, with clocks and a way to measure distance, such as a meter tape. But you take it one step further. Think of a train pulling out of the station. When the train starts up, it is moving very slowly. Before long, it is speeding down the track. To find out the train's acceleration, you need to find out how far the train has traveled at the end of each second.

You might find that the train traveled a total of 1 m at the end of 1 s, a total of 4 m at the end of 2 s, 9 m at the end of 3 s, and so on.

Is the train going at a constant speed or accelerating? To find out, we calculate its speed second by second and compare. If the speed is always the same from second to second, the speed is constant. If the speed changes from second to second, the train is accelerating.

To determine the train's average speed during any second, use the speed equation.

$$v = \frac{\Delta x}{\Delta t}$$

At the end of the first second ($t = 1$ s), the position was $x = 1$ m. The average speed during the first second was 1 m/s.

During second 2, however, the train moved 3 m more to the right ($\Delta x = 4$ m – 1 m = 3 m). Using $\Delta x = 3$ m and $\Delta t = 1$ s, the average speed during second 2 was 3 m/s.

Below is a table that shows the change of position each second and the calculated speed at the end of each second.

The speed column shows us what happened during the first 10 s after the train pulled out of the station. After 10 s, the train's position was 100 m to the right. Notice that the train covered the first half of that distance, 50 m, in a little over 7 s. It covered the second 50 m in less than 3 s.

We can calculate the train's average speed for the entire 10 s by dividing the train's ending position by the change of time.

$$v = \frac{\Delta x}{\Delta t} = \frac{100 \text{ m}}{10 \text{ s}} = 10 \text{ m/s}$$

Looking again at the table, we can see that the first 5 s of travel was slower than the average speed, and the second 5 s was faster than the average speed.

Speed was changing second by second. We can see that by looking at the speed column in the chart. But how much was it changing? What was the rate of change of speed?

Time (s)	x (m)	Δx (m)	Δt (s)	v (m/s)
0	0	-	-	-
1	1	1	1	1
2	4	3	1	3
3	9	5	1	5
4	16	7	1	7
5	25	9	1	9
6	36	11	1	11
7	49	13	1	13
8	64	15	1	15
9	81	17	1	17
10	100	19	1	19

Let's add one more column to the table. The new column will be change of speed per unit time, $\Delta v/\Delta t$.

The $\Delta v/\Delta t$ column shows a steady change of speed per second. After the train got rolling, its speed changed at the rate of 2 m/s *every second*. The change of speed, Δv, per second was the acceleration of the train. Acceleration is the rate at which the speed changes per unit time.

In this train example, position was measured in meters from a zero point, and time was measured in seconds. Thus, the speed is in units of meters per second.

With each passing second, the speed was 2 m/s faster than it was the second before. The speed was changing. But that acceleration was constant—2 m/s each second. We can say the acceleration was 2 m/s *per second*. That is usually written 2 m/s^2. We read that as 2 meters per second squared.

There is one problem with this train example. Constant acceleration of a train or other vehicle is as rare in real life as constant speed is. You will learn more about other factors that affect acceleration.

Think Questions

1. **What is the difference between speed and acceleration?**
2. **Does a flying bird move with constant speed or with acceleration? Explain.**
3. **What is an example of deceleration?**

Time (s)	x (m)	Δx (m)	Δt (s)	v (m/s)	$\Delta v/\Delta t$ (m/s^2)
0	0	-	-	-	-
1	1	1	1	1	1
2	4	3	1	3	2
3	9	5	1	5	2
4	16	7	1	7	2
5	25	9	1	9	2
6	36	11	1	11	2
7	49	13	1	13	2
8	64	15	1	15	2
9	81	17	1	17	2
10	100	19	1	19	2

Sleek high-speed trains crisscross Europe on specially designed tracks and routes at average maximum speeds of 230 to 250 km per hour. However, neither the speed nor the acceleration along these rail lines is constant.

Autumn leaves fall to the ground because of gravity. Gravity is a force that acts between all objects that have mass, pulling them toward each other.

Gravity: It's the Law

We have all heard of the law of gravity, but just what is it? What gives it the status of law?

Gravity is always there and has predictable effects on things. On Earth, gravity makes things go down. That's the law!

Did You Know?

There are four fundamental forces known in the universe: gravitational force, electromagnetic force, strong nuclear force, and weak nuclear force.

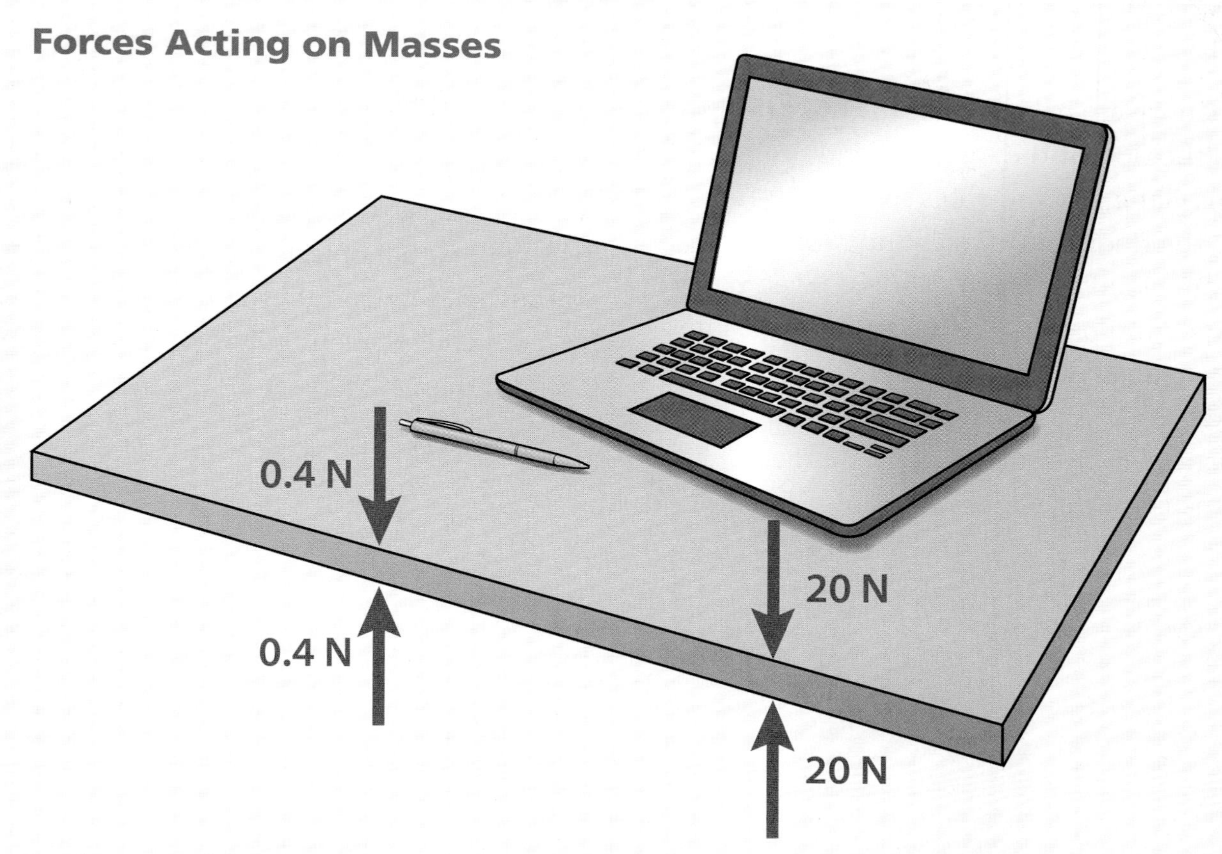

The downward force of gravity on the pen and laptop is offset by another force acting in the opposite direction: the upward force of the table. The balanced forces keep the objects in place.

How Does Gravity Work?

Gravity is a **force** of **attraction** between **masses**. Every mass, no matter how small, pulls on other masses with a **gravitational force**. But smaller masses have smaller gravitational force. On Earth's surface, gravitational forces between objects are dominated by the mass of Earth. We notice the effect of Earth's gravitational pull if our cell phone falls out of our pocket. Falling objects are pulled toward Earth's surface. Gravitational force on Earth is the force of attraction between Earth (one of the masses) and any other object.

What happens when a force acts on a mass? The mass moves. The mass will accelerate as long as force continues to act on the mass. The force of gravity accelerates any object unless an equal force opposes it. A pen sitting on a table stays in place because the table exerts a force opposed to gravity. But if you remove the table, the pen falls to the ground. Falling is acceleration toward Earth due to the attractive force of gravity.

What Else Do We Know about Gravity?

When we are babies, we learn that things fall to the floor if they are not held up. Spoons clang and tomatoes splat when they fall to the floor. Snowflakes and raindrops fall from clouds to Earth's surface. Even paper airplanes and parachutes, tossed into the air, always come down. Everything with mass is pulled down because of gravity.

Gravity has been studied and studied over the centuries. The famous philosopher Aristotle (384–322 BCE) had ideas about gravity. He thought that the **weight** of an object determines how fast it falls through air. This turned out to be wrong, but he claimed that a rock falls faster than a feather or cork because it is heavier. He also taught his students that objects move earthward at a steady speed when they are released. These inaccurate ideas about gravity were accepted for more than 1,900 years.

Every spoonful that misses the mouth will drop or drip downward. Sometimes babies test gravity by tossing food from their high chairs. Gravity always wins.

The force of gravity is hard to detect unless one of the objects—like Earth—has a lot of mass. The pull of Earth's gravity makes snowflakes fall and holds us on the ground.

Galileo's Experiments

Galileo Galilei (1564–1642) revised Aristotle's idea of gravity. He studied the motion of falling objects, but was unable to make accurate observations. Objects just fall too fast.

Then, Galileo had a brilliant idea. He reasoned that balls roll downhill for the same reason that they fall—gravitational force pulls them. The advantage of observing balls rolling down a slope is that the motion is much slower than free fall. He could accurately measure time and distance as balls rolled down ramps set at different angles.

Galileo determined that balls don't roll down ramps at constant speed. They start slow and speed up as they go. The longer they roll downhill, the faster they go. Balls accelerate as they are pulled downhill by the force of gravity.

In a series of experiments, Galileo found that the steeper he set the ramp, the greater the acceleration of the ball. He rolled balls down steeper and steeper slopes until the ramp was vertical; that is, the balls fell freely through air.

Based on these findings, Galileo revised the scientific understanding of how objects move during free fall. But he went further with his investigation of falling objects. He dropped objects of different sizes and masses from a tall building. Galileo showed that objects of different masses fall together, hitting the ground at exactly the same time.

Galileo's Ramp Trials

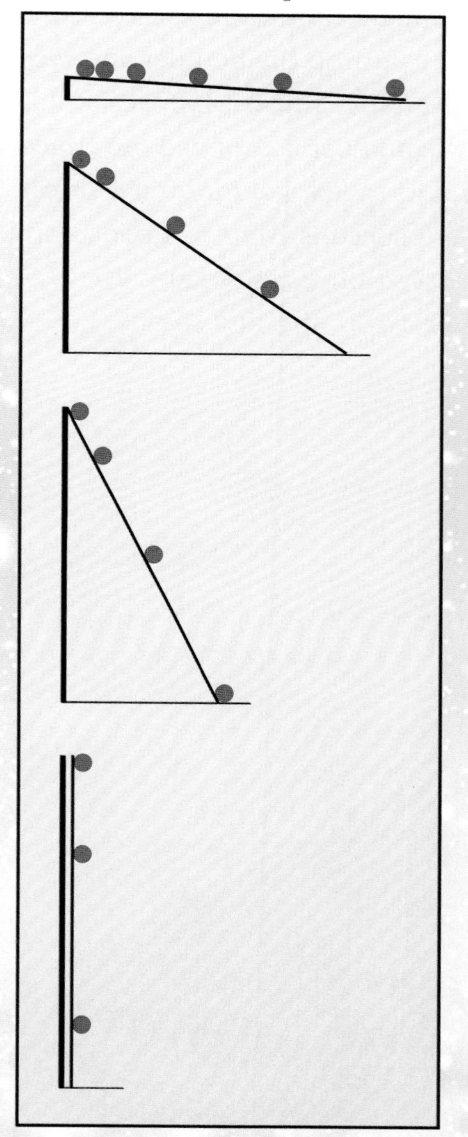

Galileo's ball-and-ramp experiments demonstrated that gravity causes falling objects to accelerate.

Acceleration Due to Gravity

The gravitational force due to Earth's mass results in an almost uniform acceleration worldwide. How much is the acceleration? To find out, we can do an experiment like the one you did in class.

Imagine a really big boulder on the edge of a cliff. The boulder has a device that measures speed, called a **speedometer**, attached to it. This speedometer can record speed second by second. We can drop the boulder and record its speed as it falls. This is what we would see.

When the boulder is dropped, no time has passed, and its speed is zero. After 1 second, the boulder's speed is 10 meters per second (m/s). After 2 seconds, its speed is 20 m/s, after 3 seconds, 30 m/s, and so on.

Time (seconds)	Speed (meters per second)
0	0
1	10
2	20
3	30
4	40
5	50
6	60
7	70

You can see a pattern in the speed and time. Every second, the boulder was falling 10 m/s faster than it was in the previous second. This change of speed is the acceleration due to gravity on Earth. The speed increases at a rate of about 10 m/s every second. Measured more precisely, the acceleration of gravity is 9.8 m/s per second. It can be written 9.8 m/s/s, which reads "9.8 meters per second per second," or 9.8 m/s^2, which reads "9.8 meters per second squared." It is OK to round that to 10 m/s^2 when your calculations do not need to be precise.

A boulder perched on a cliff is at rest. But if something jostles it off the edge, the boulder falls because of gravity. Then it moves faster as it falls farther, also because of gravity.

Acceleration Equations

We can mathematically define acceleration (a) as change in speed (Δv) over time (Δt).

$$a = \frac{\Delta v}{\Delta t}$$

If a falling object starts from **rest** ($v = 0$ m/s at time $t = 0$ s), then we can leave out the delta symbols.

$$a = \frac{v}{t}$$

We can calculate the speed of an object at any time during its fall. The equation is based on the acceleration equation. Multiply both sides by t.

$$v = a \times t$$

Here is another way of stating the relationship. For objects that start from rest, speed is equal to the acceleration times the length of time the object has been accelerating. In the case of the falling boulder, the speed is 50 m/s after 5 s.

$$v = a \times t = 10 \text{ m/s}^2 \times 5 \text{ s} = 50 \text{ m/s}$$

In the same way, we can calculate the speed after 12 s.

$$v = a \times t = 10 \text{ m/s}^2 \times 12 \text{ s} = 120 \text{ m/s}$$

How Fast Is the Fall?

Look at the first 7 s of fall. After 7 s, the boulder is traveling at a speed of 70 m/s and has fallen 245 meters (m), about the length of two and a half football fields.

How far will our boulder fall in a minute, and how fast will it be going? Using the speed equation, $v = a \times t$, we discover that it is traveling a blistering 600 m/s, which is equal to 1,342 miles per hour (mph)!

$$v = a \times t = 10 \text{ m/s}^2 \times 60 \text{ s} = 600 \text{ m/s}$$

In that minute, the boulder would travel 18 kilometers (km), assuming that it didn't hit the ground first.

Limiting Acceleration

But wait a minute! We forgot part of the system. The boulder will not get up to that speed because it is falling in air. **Air resistance** puts the brakes on the boulder, stopping its acceleration. **Friction** between the air particles and the boulder resists further acceleration. Friction is the force between surfaces that resists motion.

When the boulder stops accelerating, it stops gaining speed. Its speed is now constant. Its highest speed is the **terminal velocity**, the maximum speed an object can achieve going through the air of Earth's atmosphere.

Speed of Falling Rock

t	x	v
0 s	0 m	0 m/s
1 s	5 m	10 m/s
2 s	20 m	20 m/s
3 s	45 m	30 m/s
4 s	80 m	40 m/s
5 s	125 m	50 m/s
6 s	180 m	60 m/s
7 s	245 m	70 m/s

The falling boulder will accelerate until it reaches terminal velocity.

When these tandem skydivers reach terminal velocity, the forces of air resistance and gravity are balanced. The skydivers stop accelerating and move at a constant speed—until they open their parachute.

Skydivers reach terminal velocity a few seconds after jumping out of a plane. They can change their **surface area** to achieve different air resistance and play with their speed. Spreading out wide like a flying squirrel increases the amount of surface area they present to the air and slows their fall. Diving head first minimizes their surface area and speeds their fall. The greater the surface exposed to the air, the lower the terminal velocity.

How do skydivers return safely to Earth? They open a parachute, which has a huge surface area moving through the air. More surface area equals more friction. And that results in rapid deceleration. Their terminal velocity drops to a few meters per second, allowing the skydiver to land safely.

Acceleration is an extremely important idea in physics. We can thank Galileo for accurately describing the motion of falling (and rolling) objects. His discoveries opened the way for understanding how forces affect all objects to produce motion.

Think Questions

1. **How did Galileo figure out that a falling object accelerates?**
2. **A cat jumps from a balcony. About how fast is the cat moving after 1.5 s?**
3. **Can a falling object reach terminal velocity in outer space? Explain.**

A Weighty Matter

The doors of the elevator slide shut. It begins to glide downward, and for just a moment, you feel a little tickle in your stomach as you feel a bit lighter.

As it slows to a stop, you think you feel a downward pull. Was that just your imagination?

What you felt was a demonstration of **net force**. Net force describes all the forces acting on an object. Because the elevator only travels up and down, we can focus on the forces in those two directions.

Consider the ups and downs of elevator riding. We feel a little heavier when our elevator is accelerating upward and a little lighter when the acceleration is down. Our apparent weightiness and weightlessness are due to unbalanced forces.

Forces in Balance

When we measure the weight of an object, we are measuring the force of gravity from Earth. On Earth's surface, gravity is always pulling down with an acceleration of 9.8 meters per second squared (m/s^2). But the force of gravity is usually not the only force acting on an object.

For an object at rest, like a passenger standing in a stopped elevator, the net up and down forces on the object must be 0 **newtons (N)**. What is the upward force acting upon you to keep you from accelerating into a free fall? It is the force applied by the elevator floor. You can feel this force with your feet. When the elevator is stopped, it is pushing against you with an upward acceleration of 9.8 m/s^2. The net force interacting with you vertically is 0 N. The elevator is not moving, and neither are you.

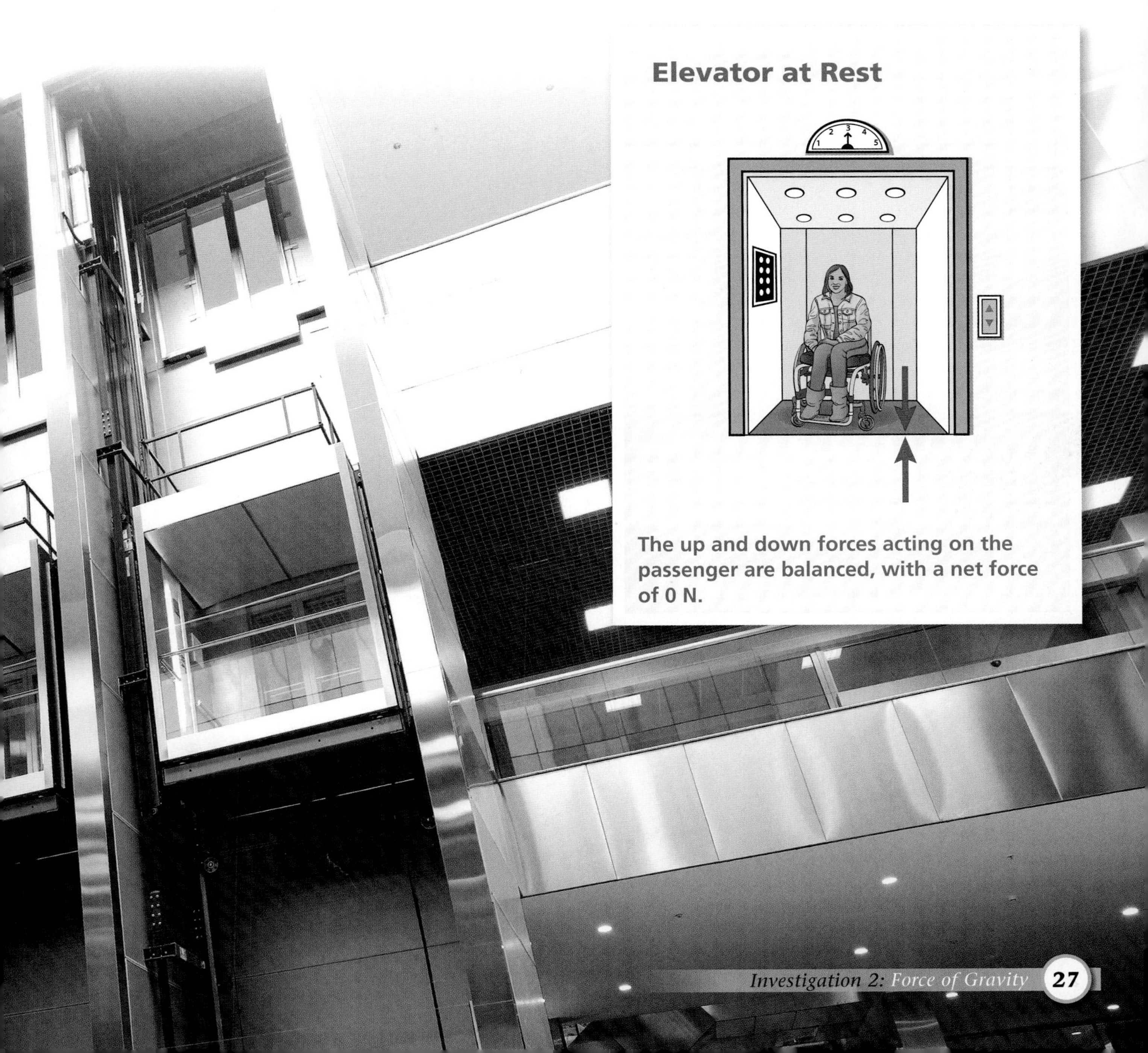

Elevator at Rest

The up and down forces acting on the passenger are balanced, with a net force of 0 N.

Acceleration and Net Force

When the elevator moves, it does not have a constant speed. It will accelerate downward, then travel at a constant speed. As it approaches the ground floor, it decelerates to a stop. Its speed changes at the beginning and end of the trip.

As the elevator accelerates downward, you feel lighter. Why is that? The elevator's acceleration changes the net force applied to your feet. It is no longer 0 N. Remember that the vertical forces acting on you were the force of gravity pulling down and the force of the elevator floor pushing up. The force of gravity does not change. This means that the force of the elevator floor pushing up on you is slightly different. It is slightly less than it was before it started accelerating downward. You are falling slightly, and you feel lighter!

Elevator Accelerating Down

As the elevator accelerates down, the upward force of the elevator floor becomes less than the downward force of gravity, and the passenger feels lighter.

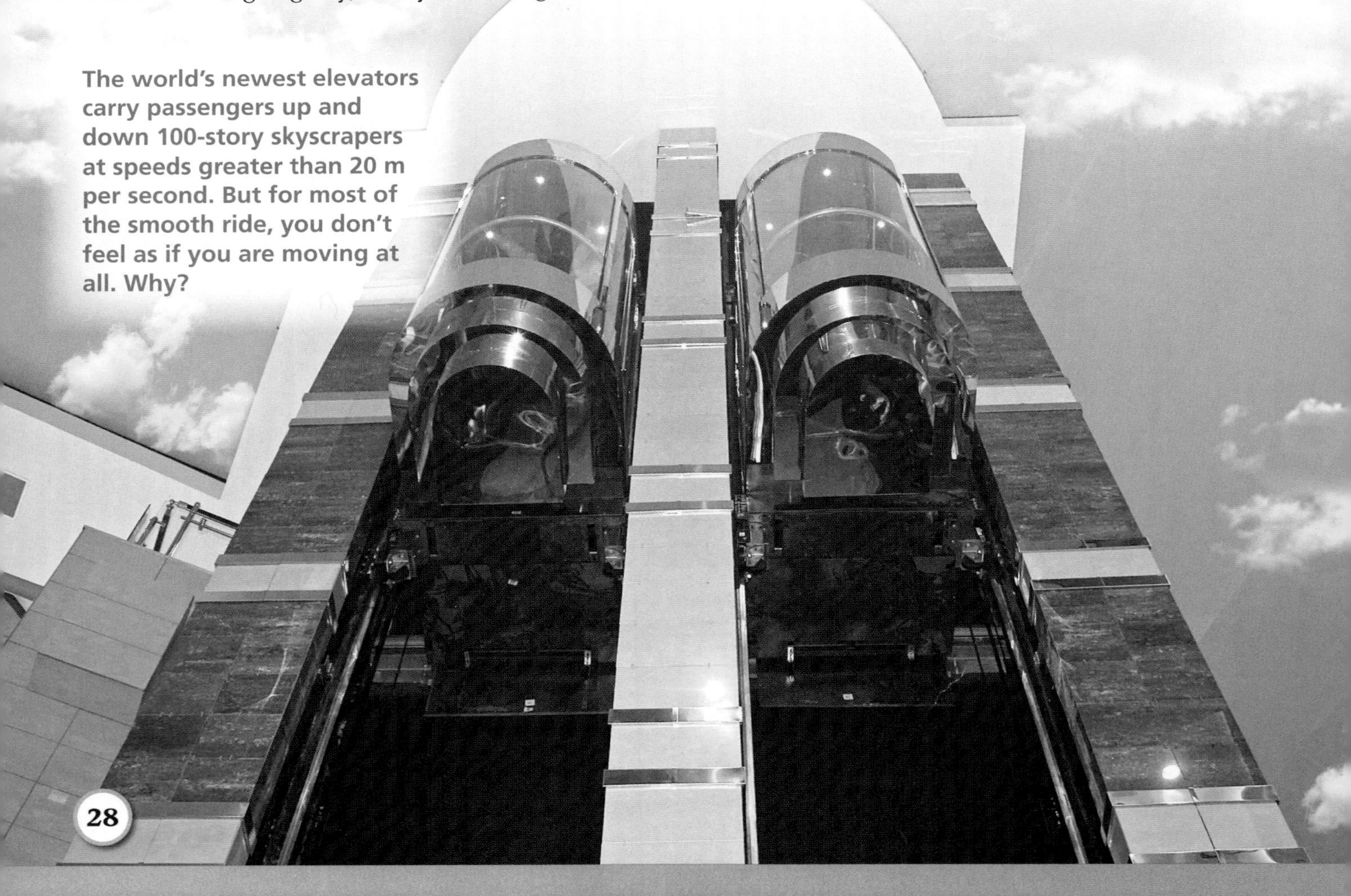

The world's newest elevators carry passengers up and down 100-story skyscrapers at speeds greater than 20 m per second. But for most of the smooth ride, you don't feel as if you are moving at all. Why?

Elevator Moving Down at Constant Speed

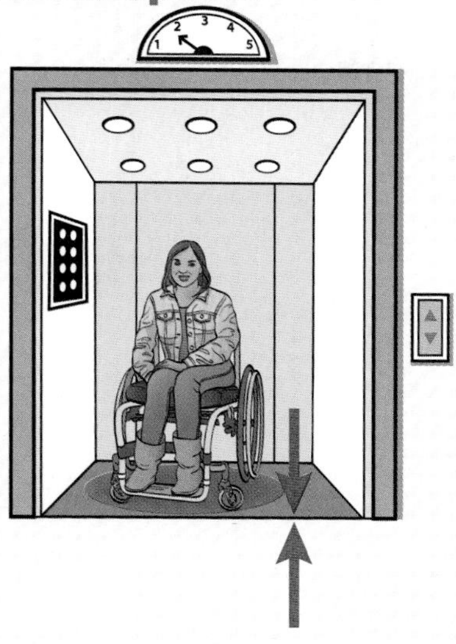

Soon, elevator and passenger drop at the same speed and the up and down forces are in balance again.

In the middle of the elevator ride, the elevator's speed and your speed are constant. There is no acceleration. You are still moving, but the upward force of the elevator equals the downward force of gravity. The forces on the passenger are balanced, meaning the net force is again 0 N, and you feel your regular weight from the force of gravity from Earth pulling you down.

Elevator Decelerating Down to a Stop

As the elevator decelerates to a stop, the upward force of the elevator floor becomes greater than the downward force of gravity, and the passenger feels heavier.

Finally, the elevator slows to a stop. As it decelerates, you feel heavier. For just a moment, you are still going as fast as you were before, and the elevator has started to slow. The elevator's negative acceleration changes your net force. It is no longer 0 N. The force of gravity does not change, but the force exerted by the elevator floor does. You feel it pushing back against you more than you did before—you feel heavier.

Take Note

A friend shows you a simple experiment in an elevator using a spring scale and a letter. Review the images and discuss with a partner how the elevator's changing acceleration could explain these results. Write a summary in your notebook.

The postal scale and elevator are at rest.

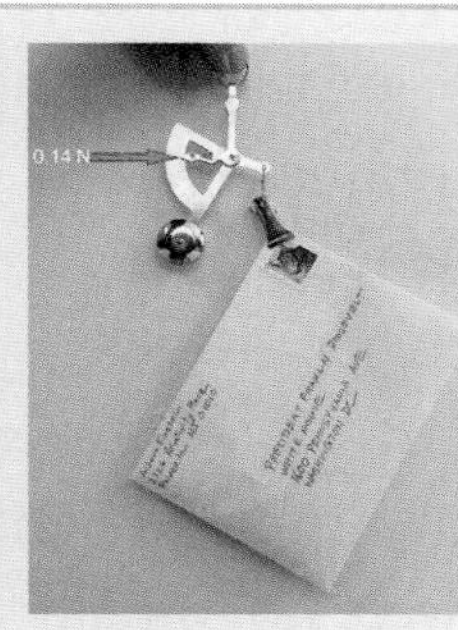

The elevator accelerates downward.

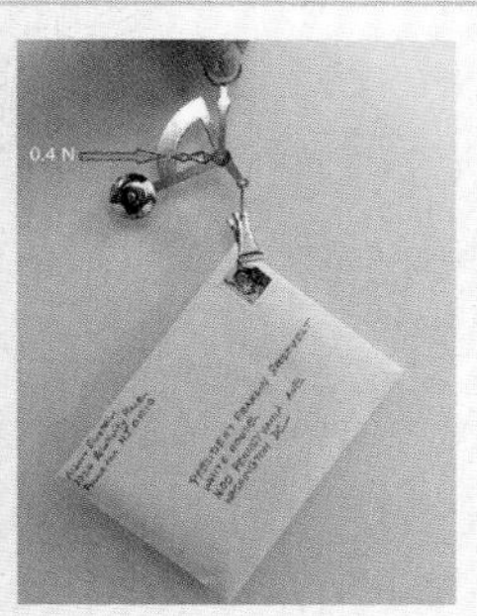

The elevator decelerates as it approaches the first floor.

Some fast-dropping rides make us feel weightless as we fall. But that weightlessness is not the absence of gravity, it's the absence of the force of your seat pushing back in the opposite direction as gravity.

Mass and Weight

When you rode in the elevator, you felt a tiny change in weight. This change in sensation of weight is exaggerated in amusement park rides. In fact, you may have felt "weightless" on a ride that accelerates suddenly toward Earth. These rides follow the same principles of physics as the elevator ride—but are much more dramatic!

When you are not on an accelerating elevator or ride, when you measure your weight, you are measuring the force of gravity from Earth pulling you down. Your weight would be different if you were not on Earth, but instead on the Moon or Neptune. These objects exert different forces of gravity, for reasons that we will explore later.

What about mass? Did your mass change in the elevator? Would your mass change on the Moon or Neptune? The answer is no. An object's mass is how much matter (stuff) is in the object. Your mass never changes during an elevator ride, nor during a ride at the amusement park.

Think Questions

1. **What is the difference between constant speed and acceleration?**
2. **Describe how you could feel weightless on Earth.**
3. **What is the difference between weight and mass?**

Gravity in Space

The surface of Mars is cold and dry. Re out before you. Wind blows dust into the air.

And with a gravitational acceleration of just 3.7 meters per second squared (m/s^2), you lightly bounce with each step you take.

Gravity and Mass

Gravity is a force of attraction between masses. The strength of the force depends on the size of the masses and the distance between them. Small masses exert very little gravitational force. The force due to gravity between two marbles is extremely weak. Compare that with the force due to gravity between Earth and the Moon, which is huge. Gravity holds the Moon in its orbit around Earth and pulls earth materials toward the Moon, affecting tides on Earth.

Mars, smaller and far less dense than Earth, has only about 11 percent as much mass as our planet. So gravity on the surface of the red planet is much lower than here at home.

we notice the effects of gravity e. When you drop your keys or trip , gravity is at work. Remember, the gth of the force due to gravity depends the mass of both objects. Large masses, like trains, ships, and buildings, are pulled toward Earth with more force than smaller masses, like coins, cherries, and ants.

Gravity in Space

The other factor that affects the force of gravity is the distance between the masses. Consider an ordinary human standing on Earth's surface. We know the acceleration of gravity on Earth's surface is 9.8 m/s^2, also known as g. This is acceleration in the direction of Earth's center of mass.

Now think about a human on the International Space Station. You may have seen videos of the astronauts floating through the air, hair sticking out in all directions. Orbiting about 400 kilometers (km) above Earth, the astronauts feel weightless. Does that mean there is no gravity up there?

Even at a distance of 384,400 km (distance not shown to scale), Earth and the Moon interact due to gravity. The Moon revolves around Earth because of the pull of Earth's gravity. Earth's ocean rises and falls in tides because of the pull of the Moon's gravity.

Understanding Orbits

To answer this question, we first must understand orbits. An orbit is the curved path of an object around another object in space. Gravity keeps the object in this path. A great physicist, Sir Isaac Newton (1643–1727), figured out how this works.

Newton thought about gravity for decades. He studied Galileo's experiments and conducted many of his own. He was the first person to describe the law of gravity as an interaction between two masses. Then, he used his understanding of gravity to explain planetary motion. He also realized that if gravity explained planetary motion, there must be gravity in space. Newton did not have computer simulations and could not collect data from space. He figured out the laws of physics using mathematical equations and logic.

On the orbiting International Space Station, anything (or anyone) that is not tied down will float. This apparent weightlessness is referred to as microgravity.

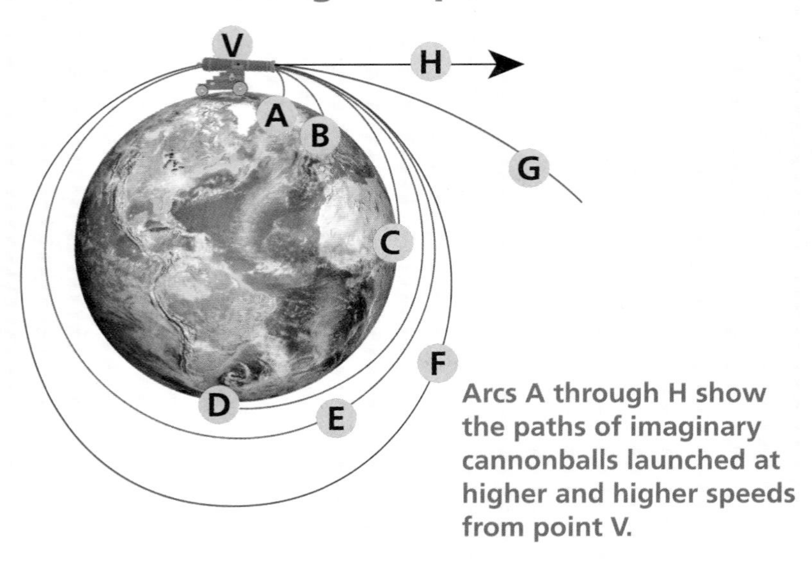

Arcs A through H show the paths of imaginary cannonballs launched at higher and higher speeds from point V.

Newton's Cannon

Newton wanted to understand how planets orbit the Sun, and how the Moon orbits Earth. He imagined a cannon that could shoot a cannonball at high speeds in a horizontal direction. A typical cannonball travels some distance through the air before landing on Earth's surface. Newton imagined what would happen if his cannon were positioned high above the surface of Earth, on an imaginary mountaintop a thousand times taller than Mount Everest.

The cannon is at position V in the diagram. It is aimed exactly parallel to the ground. If there was no force of gravity, the cannonball would follow a straight line (H) in the direction that it was fired. It would get higher and higher above the ground.

In reality, the force of gravity acts on the cannonball. It falls to the ground at some distance from the cannon. How far it travels before falling depends on its initial speed. Newton realized that the faster the cannonball left the cannon, the farther it would go before falling to Earth's surface. Examine lines A, B, C, and D in the diagram. The lines represent a cannonball shot at faster and faster speeds, traveling farther and farther.

Now, examine line E. What has happened to the path of the cannonball? It left the cannon with so much speed that it never reaches Earth's surface. It has completed an orbit around Earth. Is the cannonball still falling? Yes! During the entire orbit, the force of gravity acts on the cannonball. But the second force, the force of the cannon's blast, was exactly enough to return the cannonball to the spot it started from.

If the cannon shoots with even higher speed, what happens then? Line F is the cannonball's path with higher speed and is elliptical in shape. Still higher speeds would make the cannonball leave Earth's orbit altogether (G) when the force of the cannon blast is greater than the force of gravity pulling the cannonball down to Earth. The minimum speed required for this to occur is the **escape velocity**.

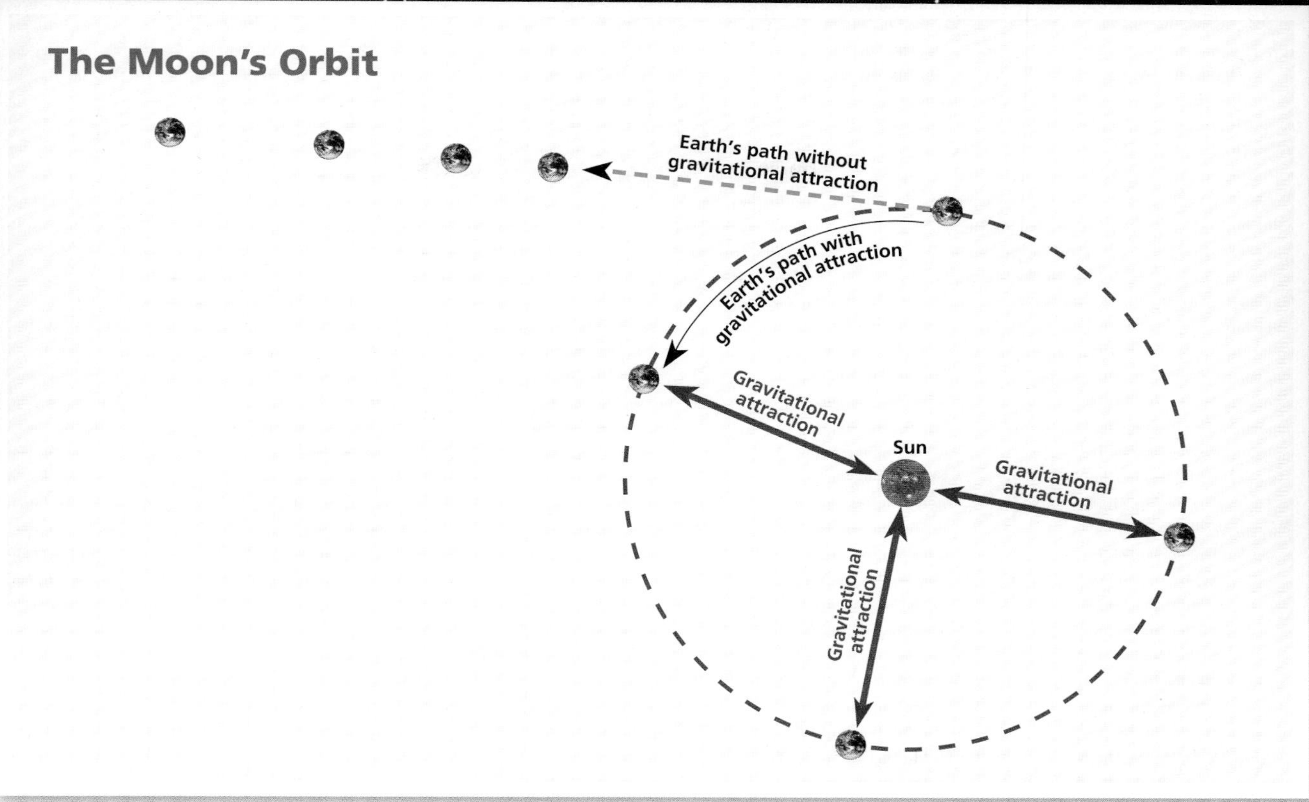

Gravitational force attracts the Moon toward Earth and keeps it in orbit.

Gravitational Force

Gravity is one of the four known fundamental forces in the universe. Others are electromagnetic force and two kinds of nuclear force. These four forces make everything in the world behave in ways we understand. Newton described the force of gravity using the term **centripetal force**. Centripetal comes from the word *centrum*, the Latin word for center.

Newton recognized that the force of gravity is always toward the center of mass. He realized that gravity is responsible for objects falling to the ground (toward the center of Earth). This gravitational force also keeps planets and satellites in orbit.

Newton also realized that an object in motion will naturally travel in a straight line forever unless a force changes its direction. The force of Earth's gravity keeps the Moon from flying off in a straight line like that in the diagram. But the force of gravity pulls the Moon toward Earth, continuously changing its direction of travel. The resulting motion is an orbit.

Imagine taking a yo-yo by the end of its string and swinging it in a circle over your head. If you let go of the string, what happens? The yo-yo stops going in a circle and flies off in a straight line. As long as you keep applying a centripetal force (pulling on the string) to change the direction of the yo-yo, it continues to orbit your hand.

The centripetal force of the string is like the gravitational force of Earth pulling on the Moon. Gravity also keeps the International Space Station in a circular orbit around Earth, and keeps Earth (and the other planets) in a circular orbit around the Sun.

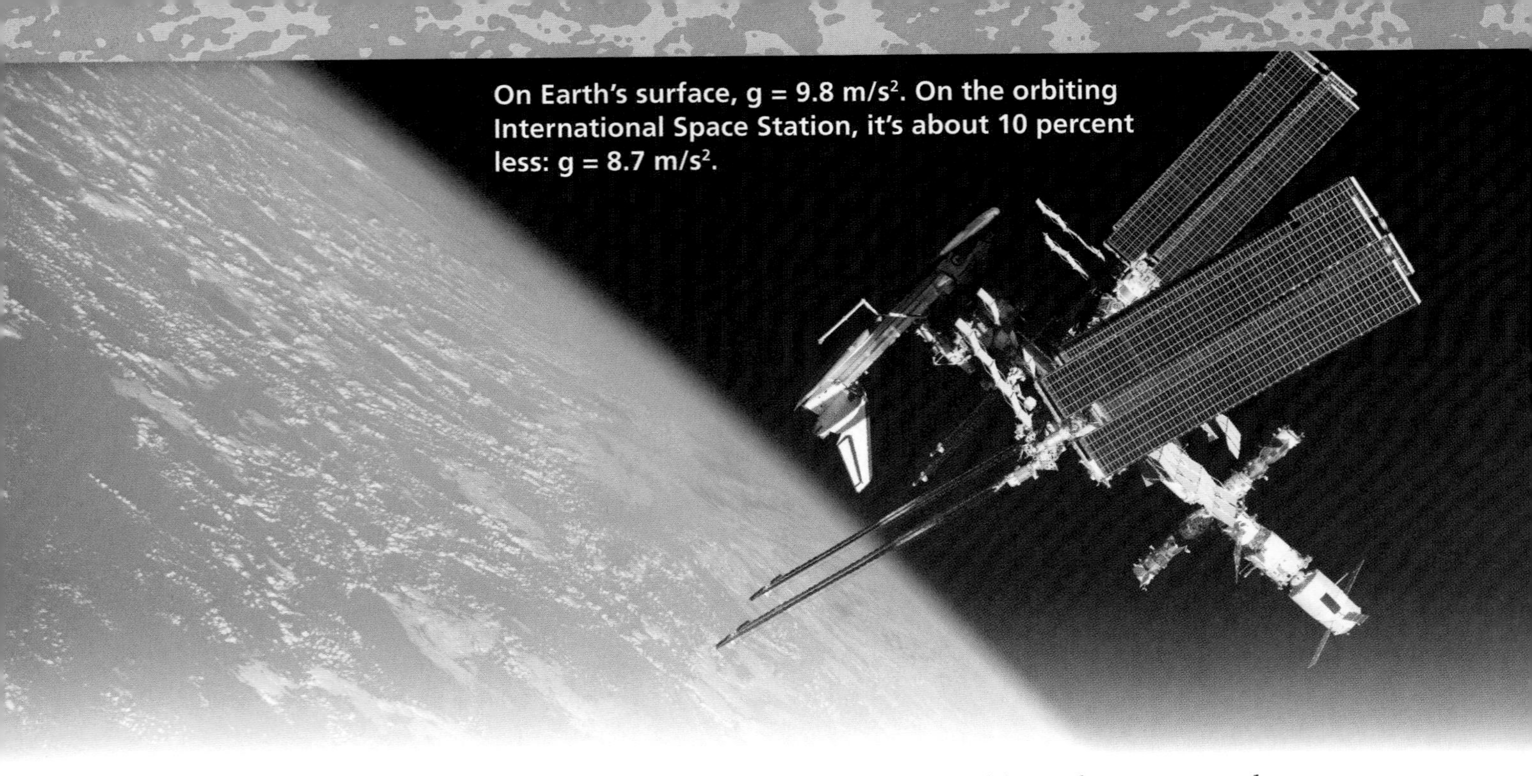
On Earth's surface, g = 9.8 m/s². On the orbiting International Space Station, it's about 10 percent less: g = 8.7 m/s².

Gravitational Fields

So, is there gravity at the International Space Station? The answer is yes, gravity is there. The astronauts feel weightless because they are in free fall, but falling *around* Earth. An object orbits Earth because of Earth's gravity.

But there is one other factor to consider. The gravitational attraction between masses decreases as the distance between them increases. The astronauts on the International Space Station are 400 km farther from Earth's center of mass than someone on Earth's surface. The acceleration of gravity there is about 90 percent of that on Earth's surface.

How far would you have to travel to escape Earth's gravity? As it turns out, you cannot escape. The force of gravity is a zone, or **field**, that surrounds a mass. As you travel farther and farther from Earth, its gravitational force becomes infinitely small, approaching (but never reaching) 0 g. Gravity's predictable force helps us understand the interaction of objects, not just on Earth, but across the universe.

Think Questions

1. **The acceleration of gravity on the surface of Mars is 3.7 m/s². What does that tell you about its mass compared with Earth?**
2. **Use Newton's cannon experiment to explain how the International Space Station orbits Earth.**
3. **What are some other examples of gravitational force keeping an object in orbit?**

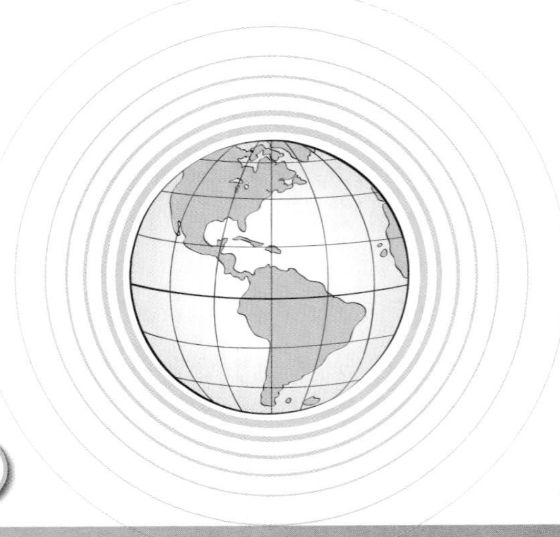
Earth's gravitational force decreases with distance.

Potential and Kinetic Energy

A giant firework explodes in the sky, with shining light and a loud boom.

A bus loaded with passengers slowly drives up a hill. You stand next to a campfire, roasting a marshmallow. What do all these things have in common?

Light, sound, motion, and heat are all evidence of energy. Energy can take many different forms. Some forms are quite noticeable, like the light energy from fireworks or the **thermal energy** from a flame. Others can be detected only if you know what to look for.

A fireworks display is an energy showcase, producing big booms, bright lights, and blazing heat. But that's not all: Fireworks shooting across the sky have kinetic energy. And chemical energy stored in the missiles and released in controlled explosions is the source of the fun.

Potential Energy

In class, you looked for evidence of **potential energy**. Potential energy is stored in the position of an object. When you lift a marble high on a ramp, you **transfer** potential energy to the marble. When you place it even higher, you transfer more potential energy.

The key to understanding energy is to think about energy transfers. Where does the energy come from? Where does it go? Energy cannot be created or destroyed. Energy that transfers to a marble must come from somewhere else in the system. In this case, the energy comes from the motion of your arm. The energy of motion is **kinetic energy**. You transferred energy to the marble when you changed its position. You increased its potential energy.

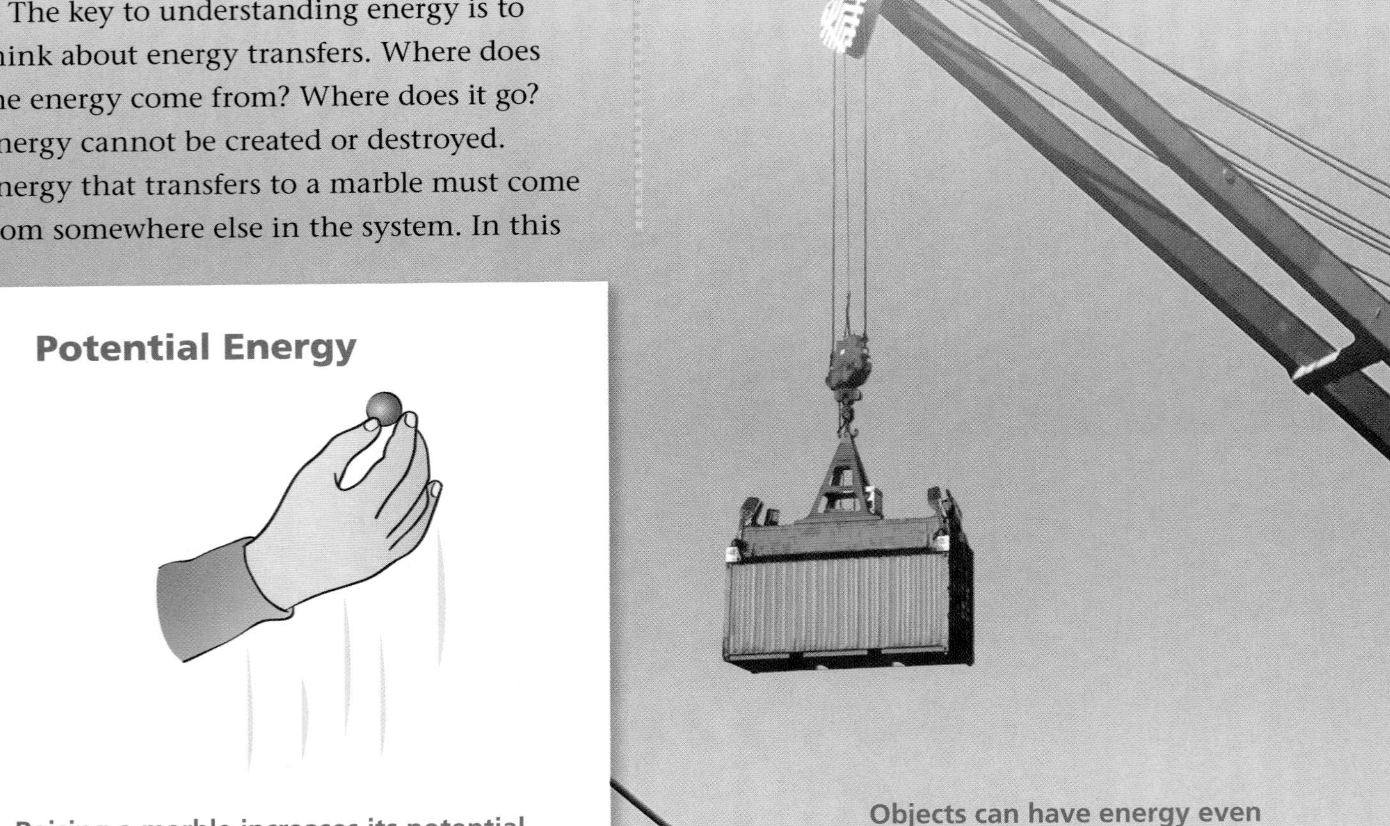

Potential Energy

Raising a marble increases its potential energy due to position.

Objects can have energy even if they are not in motion. This truck-sized container suspended by a crane has potential energy because of its height.

Take Note

Energy cannot be created, only transferred. Where did your body (arm) get the energy it transferred to the marble?

Release the marble. It rolls down the ramp. The marble accelerates as the force of gravity pulls on it. As the speed increases, the marble's kinetic energy increases. Meanwhile, its potential energy decreases, because its position is moving lower on the ramp. In this part of the system, potential energy transfers to kinetic energy. The higher the marble is on the ramp to start, the more potential energy it has. It can then gain more kinetic energy when released.

What other **variables** can you can change to increase kinetic energy? You can increase the mass of the object on the ramp. As you know, the force of gravity is an interaction between masses. When you increase the mass of one object, there is greater attraction. A more massive marble (or a group of marbles) will have more gravitational force. We can use Newton's second law to quantify that increase of force.

$$F = ma$$

On Earth's surface, acceleration (a) is a constant, g, 9.8 meters per second squared (m/s^2). If you increase the mass (m), what happens to force (F)? It must also increase. So a more massive marble on the ramp has greater potential energy because the gravitational force pulling it down is greater. The more massive marble will then gain more kinetic energy as it accelerates down the ramp.

Changing Variables

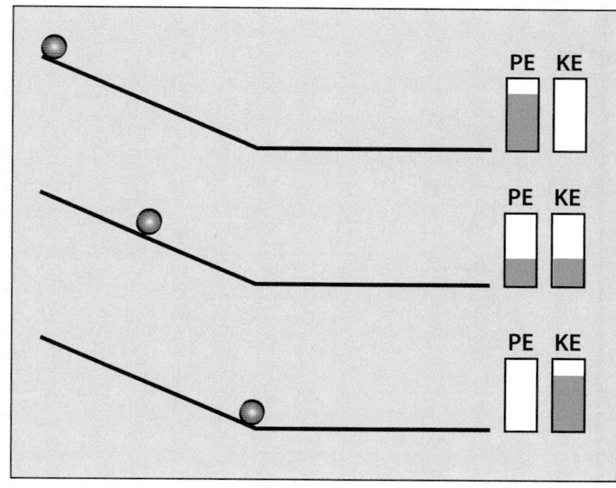

Position: As the ball rolls down the ramp, its potential energy transfers to kinetic energy.

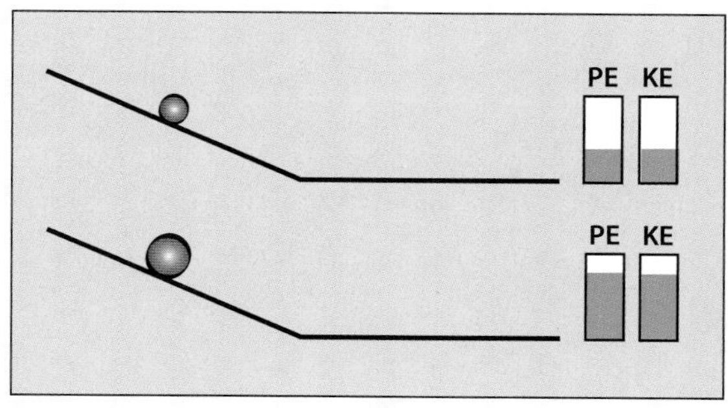

Mass: The greater the mass, the greater the potential and kinetic energy.

Your marble reaches the end of the ramp. It slams into a carefully placed cube. The cube slides across the table (evidence of kinetic energy), then stops. Where did the cube get its kinetic energy? It transferred from the marble's kinetic energy during the moment of **collision**. Our evidence for this transfer is (1) the cube started moving, and (2) the marble slowed down. We can **infer** that the marble transferred kinetic energy to the cube upon impact.

Did the cube keep moving forever? No, it slid to a stop. We know that its kinetic energy cannot have been destroyed. That energy transferred to another part of the system. Friction is responsible for this energy transfer. Friction is a force that resists motion between surfaces. As the force of friction acted on the cube and table, kinetic energy transferred to the environment as thermal energy. Both surfaces heated up a tiny amount as friction caused the cube to stop.

Think Questions

1. **What are two ways in which you can increase the potential energy of a marble on a ramp?**
2. **What happens to energy in a collision between objects?**
3. **What causes an object sliding on a surface to stop?**

Energy Transfer in a Collision

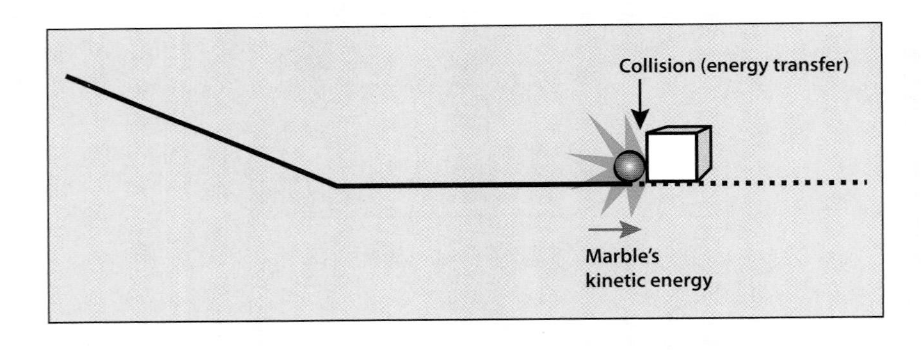

The marble transfers energy to the cube.

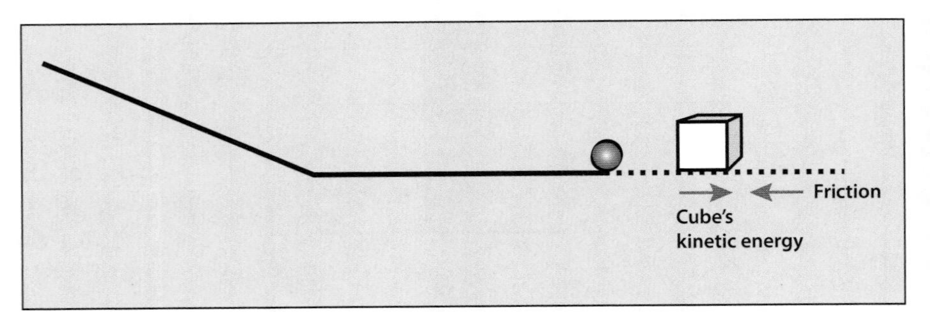

Friction slows the cube to a stop.

Imagine the consequences if the driver is not paying attention to the road. Stopping a vehicle involves good brakes and tires, driver alertness and reflexes, and factors including speed and distance.

Avoiding Collisions

Driving at 32 kilometers (km) per hour (20 miles per hour [mph]) through a neighborhood, a driver sees a soccer ball bounce into the street ahead.

The driver slams on the brakes and stops just as a small child runs out after the ball. A tragedy is avoided.

What does it take to stop a car? Simple, the brakes stop it. But what goes into the simple act of stopping a car?

Stopping Distance

When the driver slams on the brakes, an average midsize car traveling 32 km per hour (20 mph) needs about 9 meters (m) (30 feet) to stop. But even before this, the driver has to think about stopping and move a foot from the gas pedal to the brake. At 32 km per hour (20 mph), during that time the car moves about 9 m (30 feet) more. That is a total of 18 m (60 feet) to come to a full stop.

Stopping Distance, Slow Speed

32 km per hour (20 mph)

Even at slow speeds, stopping distance is many car lengths.

Stopping Distance, Fast Speed

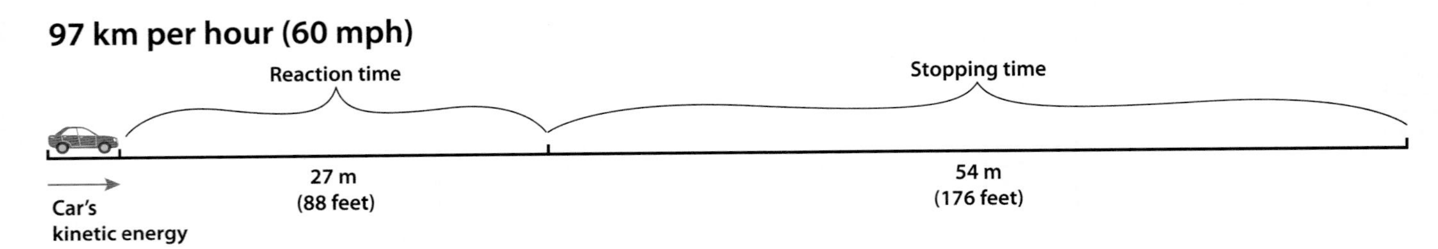

The greater a car's speed, the greater the time and distance it takes to stop. That's why it's important not to "tailgate," or travel too closely behind another vehicle.

A car traveling at 97 km per hour (60 mph) takes longer to stop, and travels farther as the driver reacts. The most attentive driver might take about 1 second (s) to recognize a hazard and move a foot to the brake. Then, the car takes about 2 s to stop. During this reaction time, the car has traveled about 80 m (264 feet)!

Any distraction increases the driver's response time. Drivers are safest when they are alert, sober, and paying attention to the road. Stopping distance relies on friction between the car's tires and the road. Wet or icy conditions will decrease friction, and increase stopping distance.

Did You Know?

Looking at a text message can take 3 to 4 s. A car moving at 97 km per hour (60 mph) will travel the length of a football field during that time.

Energy Transfer

What else happens if a car is driving faster? You learned in class that increasing the speed of a moving object increases the energy transferred in a collision by the increase in speed *squared*. Energy is often measured in **joules (j)** or kilojoules. Increasing the speed of the car from 32 to 97 km per hour (20 to 60 mph) increases the speed by a factor of 3. That same increase of speed increases its kinetic energy by 3^2, a factor of 9.

The kinetic energy also increases as the *mass* increases. The kinetic energy of a semitrailer truck traveling the same speed as a sports car is much greater. It will also require more stopping distance.

The force of a collision transfers energy between objects when they collide. If there is not enough room for a vehicle to safely stop, a collision may occur. If your car is hit by a slow, lightweight car, less energy will transfer. Everyone may escape the accident unharmed. But if the car is speeding or more massive, more energy transfers during the collision. The result of all this energy transfer can be property damage or bodily harm.

In a collision between a car and a pedestrian or bicyclist, the difference in mass is significant. Even a small car is far more massive than the person or bike. That plus the fact that the pedestrian or bicyclist has very little protection compared with a car means that a collision is even more likely to result in serious injury.

The speed limit for a road depends on road conditions. Urban interstate highways are typically fairly straight, and most allow traffic at 80 km per hour (50 mph) or higher. Winding roads tend to have lower speed limits.

Staying Safe in Traffic

Everyone wants to avoid an accident. Safe driving practices and government regulations like speed limits can make driving safer.

When following another car, a driver should always keep a safe distance behind other cars. What is a safe distance? One easy rule of thumb is to keep one car length between vehicles for every 10 mph (16 km per hour). So if you are traveling 30 mph (48 km per hour), there should be three car lengths between the cars.

Speed limits are set by state and local agencies. They judge what is safe for that location. A curvy road should have a lower speed limit, because the driver's line of sight is limited, and he may not have time to react and stop safely. Lower speed limits are usually posted in neighborhoods, especially around schools. The highest speed limits are found on limited-access highways in rural areas. Some highways have a different, slower speed for trucks or a slower speed at night.

Take Note

Why might a highway have a slower speed for trucks? Think about kinetic energy as you respond.

Think Questions

1. **What happens to the stopping distance of a sleepy or distracted driver?**
2. **Why should you reduce your speed on an icy road?**
3. **How much more kinetic energy does your car have when it changes its speed from 16 km per hour (10 mph) to 80 km per hour (50 mph)?**

Newton's Laws

Sir Isaac Newton (1643–1727) was one of the greatest scientific minds of all time.

You have learned about Newton's experiments to explore gravity, and his mathematical models of the movement of objects in the solar system. He also built telescopes and studied light. Perhaps his most famous work is known as Newton's laws of motion.

Newton explained that the motion of objects follows three basic principles. His three laws of motion help us understand and predict how forces affect objects on Earth and throughout the universe. (Solar system objects and distances not shown to scale.)

Newton's Early Life

In 1643, Isaac Newton was born in England. He lived with his grandmother on a small farm in the countryside. Throughout his childhood, he was rather frail and timid. He liked to read and think about things. This interest would lead Newton to tackle some of the biggest scientific questions of his time.

Newton did very well in school and studied mathematics at the University of Cambridge. There, he designed equations that became the foundations of calculus. In 1665, he received his degree just as the country was hit by the black plague. The University of Cambridge closed. People who could leave the infected cities did so. Newton returned to the countryside where he was born. He had the freedom to work on the ideas that interested him.

At the age of 22, Newton had a breakthrough about gravity. He wondered why an apple falls down, rather than in some other direction. He reasoned that it fell down because Earth pulled on the apple, and at the same time, the apple pulled on Earth. Gravity, thought Newton, is an attraction between two masses pulling on each other. He was able to apply this idea, called the universal law of gravitation, to explain why apples fall to Earth and why planets orbit stars.

Gravitational Force

Newton's observation of falling apples helped him realize that Earth pulls on the apple and the apple pulls on Earth. He then identified gravity as the fundamental force controlling the motion of objects in space.

Laws of Motion

Newton was aware of Galileo's work on motion. He spent many years expanding on Galileo's ideas about force and motion. In 1687, Newton's most important scientific publication was released, *Mathematical Principles of Natural Philosophy*. This book described Newton's three laws of motion.

Law 1. Every object stays at rest unless an outside force acts on it. If an object is moving, it travels in a straight line at a constant speed unless an outside force acts on it.

Law 2. The acceleration of an object is directly proportional to the force on the object. It is inversely proportional to its mass ($a = F/m$, or $F = ma$).

Law 3. For every action or force, there is an equal and opposite reaction or force.

These laws are such a part of today's scientific knowledge that it is hard to imagine a time before we understood them. Newton was the one who pondered, experimented, and analyzed all of his findings to develop these three important laws.

Applying Newton's Laws

How do these three laws shape our thinking? We can explain almost all motion with these laws. For example, a basketball player has the ball. He wants to move a little closer to the net before taking his shot, so he runs down the court, dribbling the ball.

Think about law 1. What outside force may cause the ball to start or stop moving? Think about law 2. How would the game be different if a bowling ball was used? Think about law 3. What forces are at work as the players run?

Applying Newton's Laws

The force and motion of the player and the ball as he runs and dribbles down the court are explained by Newton's laws.

At each step, his foot touches the ground. He applies a force to the ground, and the ground applies a force back (law 3). Otherwise he would have no way to move forward. His shoes are designed to have enough friction so that he can quickly change direction (law 1). To dribble the ball, he pushes it down with his hand (law 1). The ball is pulled to the ground by the force of gravity (law 1), hits the ground with a force, and then bounces back from the force of impact (law 1). The player knows how gravity will affect the ball (law 2), because it has a known mass.

He decides to take the shot. He lifts the ball above his head, and pushes the ball (law 1) toward the basket. He knows exactly how much force to use (law 2) to accelerate the ball. It arcs up through the air, slowing slightly from air resistance (law 1). The ball arcs down as the force of gravity (law 1) accelerates it toward Earth (law 2). The ball enters the net and swishes through. The net swings to the side with the ball (law 1) and then back in the other direction (law 3).

Basketball Shot

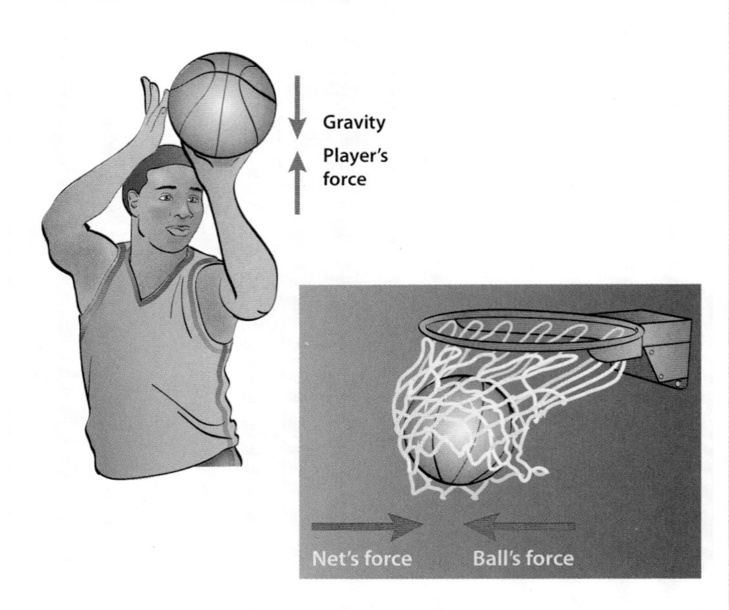

All three laws of motion describe the forces at work in scoring a basket.

The Large Hadron Collider, near Geneva, Switzerland, is the largest experimental facility ever built. It tests properties of light and matter in an underground track that is 27 km around.

Advances in Physics

Many physicists have refined and improved upon Newton's work. Scientists in the early 1900s began to understand more about light and other forms of radiant energy, which travel at the fastest known speed. They found that Newton's laws of physics do not explain phenomena at near **light speed**. In 1905, Albert Einstein (1879–1955) proposed the **theory of special relativity**. It helps explain the relationship between space and time.

The theory of special relativity is now considered the most accurate model of motion. Scientists continue to test light and other radiation in large-scale physics experiments. The findings from these experiments agree with special relativity. And other physics theories have helped us understand gravity and the motion of masses smaller than atoms. Nevertheless, Newton's laws of motion still accurately explain most everyday movements of objects.

Think Questions

1. **To what fields of science did Newton contribute?**
2. **Write your own example that includes Newton's three laws of motion.**

Engineering a Safer Car

When you pass an accident on the highway, it can be scary to see the broken glass and rescue vehicles. You hope that everyone is OK.

Unfortunately, over 2 million people are injured and over 30,000 people die in traffic accidents in the United States each year.

The first automobiles transported people and cargo easily, but slowly. By the early 1900s, new designs allowed cars to travel faster. This presented new safety concerns. More speed transfers more kinetic energy in a collision and creates more damage to metal, glass, and passengers.

Consider a car hitting a wall at 50 kilometers (km) per hour (30 miles per hour [mph]). The energy transferred in a collision depends on the speed and mass of the objects involved. If the car has a mass of 1,200 kilograms (kg), it hits the wall with a lot of force. As we know from Newton's third law, the wall pushes back with an equal force.

Force between Wall and Car

The car slams into the wall with a powerful force. The wall pushes back with an equally strong force in the opposite direction.

The resulting damages depend on several factors. One factor is car design. **Engineers** work hard to design vehicles that meet an important **criterion**: to protect passengers during a collision.

Crumple Zones

A passenger's body moves at the same speed as the vehicle. When the vehicle suddenly stops, the passenger's body suddenly stops as well. This jarring change of motion damages the body's internal organs, especially the brain. **Impulse** describes a force applied over a period of time. The more quickly energy transfers in a collision, the more damage is likely to the colliding objects and passengers. One way to protect passengers is to lengthen the stopping time, which decreases the force applied to passengers.

Modern cars have **crumple zones**. These zones absorb energy during impact and lengthen the duration of the collision so the passengers do not stop as abruptly.

A safety compartment around the passenger area keeps it from changing shape. Special structures in the front and rear of the vehicle absorb the force of impact by bending and folding in specific patterns. The longer it takes these structures to crumple, the less jarring the crash is for passengers.

Sometimes a car does not look too bad after an accident. But it can still be declared a total loss if the crumple zone was damaged, because passengers might not be protected in a future collision.

Side and back car windows are made of tempered safety glass designed to shatter into tiny, harmless glass balls, not sharp splinters.

Safety Glass and Laminated Windshields

Early windshields were made of ordinary window glass. In a collision, these early windshields could break into sharp pieces. The flying pieces of sharp glass were a serious danger for passengers.

In 1919, cars started to have windshields made of glass coated with plastic. The plastic held the broken glass together in an accident. A weblike cracking pattern protected passengers from flying pieces of sharp glass. Today's windshields are still laminated and have tougher glass. Their design keeps them from being dangerous when they break.

Windshields are composed of a strong plastic inner core sandwiched on either side by shatterproof glass. The glass may chip or crack, but the window's structure limits damage to the outside layers.

Air bags deploy automatically within milliseconds of a collision. They protect passengers by slowing their forward motion and preventing them from hitting anything hard.

Air Bags

The front of a car is still a dangerous place for a passenger during a collision. In 1941, engineers created a fabric safety bag, called an air bag. The air bag puts an energy-absorbing surface between the passenger and the front of the car. The air bag inflates and deflates rapidly during a collision. It lengthens the passenger's stopping time, helping prevent injuries.

Since 1998, all cars and light trucks sold in the United States have air bags on both sides of the front seat. Today, most vehicles have air bags on the sides of the rear seat as well. Air bags work with seat belts to protect passengers.

Did You Know?

An air bag opens at speeds close to 320 km per hour (200 mph). It can harm passengers who are too small to sit in its path or who are not wearing a seat belt. This is one reason that small children should never sit in the front seat of a car.

Seat Belts

Perhaps the most important vehicle safety feature is the seat belt. Seat belts keep passengers in the vehicle and positioned correctly for maximum protection from air bags and crumple zones.

Remember that in a car moving 50 km per hour (30 mph), a passenger is also moving at that same rate. When the car stops suddenly, an unbelted passenger keeps moving forward at 50 km per hour and can slam into or through the windshield.

The National Highway Traffic Safety Administration (NHTSA) estimates that 75 percent of passengers ejected from a car during a crash do not survive. Seat belts keep passengers in the car during a crash. The NHTSA estimates that seat belts reduce serious crash-related injuries and deaths by about 50 percent!

In spite of laws requiring use of seat belts, some people still do not wear them. Many modern cars remind the driver and front passenger to use their seat belts.

Seat Belt Protection

With Seat Belt	Without Seat Belt
 Car: 50 km per hour (30 mph) Passenger: 50 km per hour (30 mph)	 Car: 50 km per hour (30 mph) Passenger: 50 km per hour (30 mph)
 Car: Sudden stop to 0 km per hour (30 mph) Passenger: Sudden stop to 0 km per hour (30 mph)	 Car: Sudden stop to 0 km per hour (30 mph) Passenger: 50 km per hour (30 mph)

Remember Newton's first law: an object (or driver) will keep moving in the same direction until an outside force acts on it. A seat belt applies that life-saving outside force.

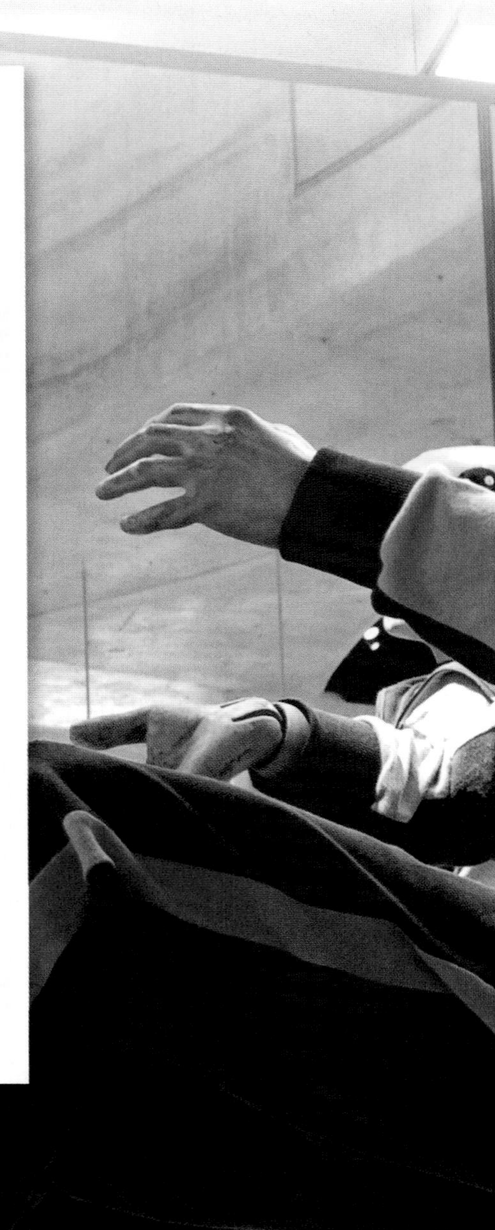

Crash Testing

Do engineers wait for a collision to study the effects on passengers and car structure? They used to. In the 1950s, however, scientists and engineers began collecting data using crash-test dummies. The data from these controlled experiments gave valuable information about the forces of a collision. Engineers used this information to create designs that limit a person's contact with the vehicle and to control the passenger's rate of deceleration.

Did You Know?

Internal injuries occur in collisions because of the laws of motion. Just as a passenger is in motion within a vehicle, so are the organs within their body. When a person's skeleton comes to a sudden stop, their organs can collide with the inside of their skeleton, causing serious damage. This is particularly true for the brain as it hits the inside of the skull, potentially causing brain injury.

Crash-test dummies are built from materials that mimic human anatomy and provide digital injury data from around 20 points on the body. The results help automotive engineers design safety improvements.

Crash Avoidance

Some vehicle designs aim to avoid collisions altogether. Better-designed brakes, steering, and tires have helped reduce collisions. Designs that allow better driving visibility have helped, too.

Engineers have also made cars smarter. Voice navigational systems keep the driver's eyes on the road, not on a map. Recent technology can tell whether a driver is not paying attention and alert her to focus on the road. What if we could remove human error entirely? Self-driving cars can safely move from one place to another without help from a human driver. The traveler indicates the destination, and the autonomous car does the driving. Without the usual human distractions and dangerous behaviors, automobile transportation could become much safer for everyone.

Did You Know?

A driver using any kind of mobile phone is much more likely to have an accident. The driver is less focused on the road and traffic conditions. Reaction time increases. Having a conversation with passengers can also increase the risk of an accident, but not as much as using a mobile phone.

Think Questions

1. **What is the relationship between increased speed of a car and energy in a collision?**
2. **Why are seat belts considered the most important safety feature of a car?**
3. **Why does talking on a mobile phone while driving increase accident rates?**

Self-driving cars are able to sense the environment and navigate without human input by using laser rangefinders, GPS, cameras, and other technology. Several automakers are piloting test vehicles.

High school athletes suffer thousands of concussions a year, most often in football, ice hockey, and soccer. Repeated head impacts can lead to long-term brain damage.

Collisions and Concussions

Have you ever bumped your head? A blow to the head can injure your brain and temporarily or permanently change the way your brain works.

If that happens, the brain injury is called a traumatic brain injury (TBI). A mild TBI is called a **concussion**.

Brain Collisions

Here's how it works. The bones of your skull surround and protect your brain. But the brain does not fill the skull entirely. Tissue and fluid fill the extra space and help cushion the brain from bumps.

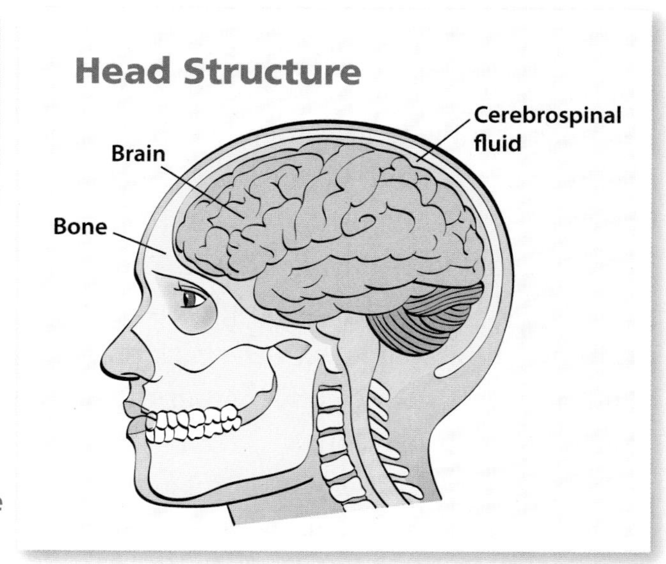

Your brain is enclosed in a helmet-like skull and layers of shock-absorbing tissues and fluids.

Brain Collisions

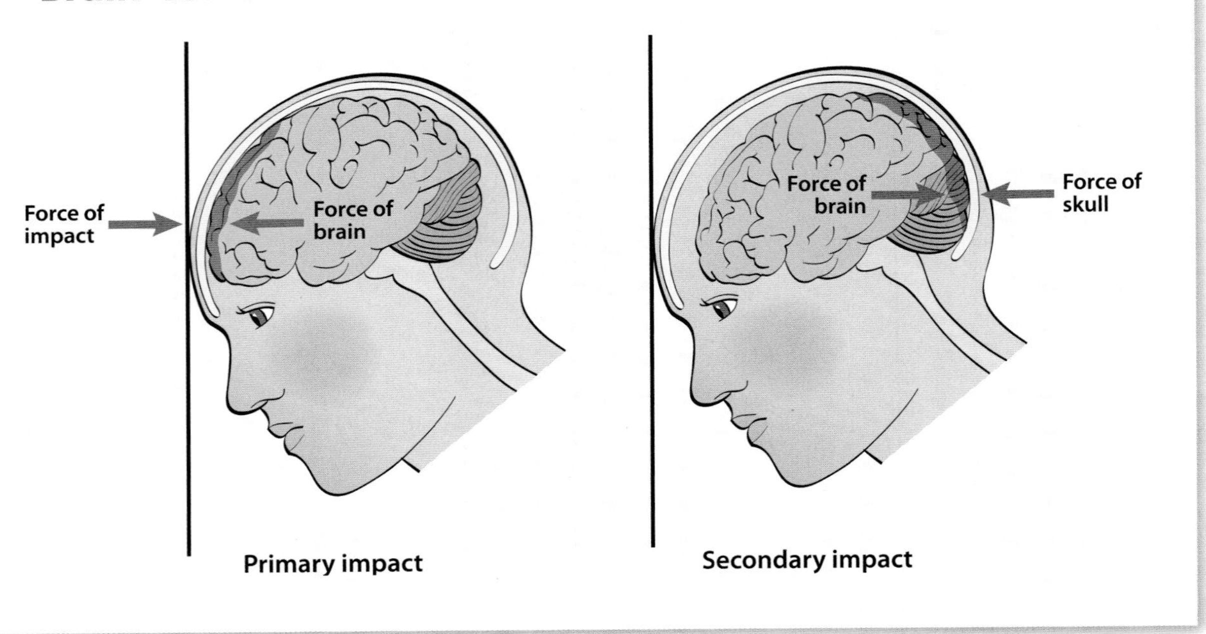

The kinetic energy of an abrupt impact is transferred through bone to the brain, which collides with the inside of the skull. In a strong collision, the brain bounces backward and collides with the opposite wall of the skull. One accident can result in multiple traumas to the brain.

When the head takes a hard hit, the force can cause the brain to move and hit the inside walls of the skull, sometimes more than once. These collisions happen in car or bicycle accidents, sports, and even minor falls.

Why does the brain hit the skull more than once? If you are moving forward and you fall, you come to a sudden stop. But your brain keeps moving forward and collides with the front of the skull. Because of Newton's third law, the skull pushes back on the brain with an equal force during impact. The brain bounces backward. With enough force, the brain may now collide with the back of the skull. You were in one accident, but the brain was in multiple collisions.

The Centers for Disease Control and Prevention estimates that about 2.5 million emergency room visits, hospitalizations, or deaths are related to TBIs every year. For teens, concussions are the largest cause of emergency room visits. Every year, about 470,000 children ages 0–14 visit an emergency room for a concussion.

Did You Know?

A brain injury can take 1 to 3 days to emerge. You should see a doctor if any of these symptoms appear after a blow to the head.

- **Loss of consciousness**
- **Severe headache that gets worse**
- **Trouble walking, blurred vision, or prolonged dizziness**
- **Confusion, slurred speech, or loss of memory**

After a Concussion

Young people with a concussion almost always heal quickly if they do not have another concussion soon after. But many people do not take enough time off from the activity that caused the first concussion. Scientists and doctors keep learning more about the long-term effects of multiple concussions. Teens' brains are particularly at risk, because they are still developing.

Bird Brains and Ram Horns

Scientists and engineers often look to the natural world for solutions to problems. Let's look at two animals who seem unaffected by frequent head impacts. Scientists have studied the anatomy of woodpeckers and bighorn sheep to see what clues they can offer for protecting human brains.

Most woodpeckers pound their beaks into the wood of trees an average of 12,000 times per day, at a speed around 7 meters per second!

Male bighorn sheep bash their heads together in rituals that involve collisions at speeds of 30 to 60 km per hour.

Why don't these animals have brain damage? Here are some important findings.

- **Size matters.** The average woodpecker brain has a mass of 2 **grams (g)**. In comparison, the average human brain has a mass of 1,400 g. With a larger mass comes more force and potential for damage.
- **A snug fit helps.** The brains of woodpeckers and rams fit more snugly in their skulls than human brains do. This snug fit reduces the back-and-forth motion and collisions between the brain and the skull wall.
- **Use bubble wrap.** Both woodpeckers and rams have tissues between their brains and skulls that cushion like bubble wrap.
- **Absorb the shock.** Woodpecker beaks and ram horns absorb much of the force of impact before it gets to the skull. These structures minimize the collision impact on the brain.

Did You Know?

Second impact syndrome (SIS) can occur when a person recovering from a blow suffers a second blow to the head. SIS can cause brain swelling that leads to serious problems. Almost all people affected by SIS are less than 18 years old.

Helmets to Protect the Brain

One simple way to help protect your brain during dangerous activities is to wear a protective helmet. In a fall, the force of gravity pulls the body down, causing a collision with the ground. Other activities can cause collisions with people or objects. It takes just a few seconds to put on a helmet. In a collision, it could prevent death or injury. Basic helmets use insulating materials to absorb force, a hard shell to protect the inner contents (both the insulation and the head), and a snug fit to ensure no back-and-forth collisions inside the helmet. A **constraint** for engineers who design helmets is to keep the helmets lightweight and comfortable, so that people will want to wear them.

Did You Know?

Some US states do not require motorcyclists to wear helmets. In these states, ten times more motorcyclists died in road accidents in 2014 than in states that require helmets.

Take Note

The main challenge with helmets for children ages 5 to 7 is getting them to want to wear one. How could you help persuade younger children that wearing a helmet is important?

Some helmet designers have improved on the basics. Several years ago, Anirudha Surabhi was in a bike accident. The helmet he was wearing cracked from the impact, and he ended up in the hospital with a concussion. Inspired by the skull structure of woodpeckers, Surabhi designed a superstrong, light helmet. These helmets can absorb three times as much force as a standard bike helmet and are 15 percent lighter.

Claire Longcroft's father, an avid cyclist, was in a bike crash. He suffered for years from headaches and memory loss. In 11th grade, Longcroft invented a liner for the helmet to provide more protection. Like the spongy tissue that separates a woodpecker's beak and skull, the liner absorbs some of the impact between the helmet and the skull. A standard bike helmet can absorb up to 88 percent of the energy of an impact. A helmet with the liner can absorb about 97 percent of the impact energy.

In Sweden, many people commute to work on bicycles. Two university students decided to design a helmet that people would want to wear. Their design sits around the neck like a collar. It is a folded air bag that inflates automatically when the wearer is in an accident.

Think Questions

1. **Why do scientists study the natural world to find solutions to human problems?**
2. **Why are humans more susceptible to concussions than bighorn sheep and woodpeckers?**
3. **What activities do you do where it would be a good idea to wear a helmet?**

Hard on the outside and squishy on the inside, bicycle helmets combine the right materials with the right design to prevent injuries and save lives.

Images and Data

Images and Data Table of Contents

Falling Ball Images

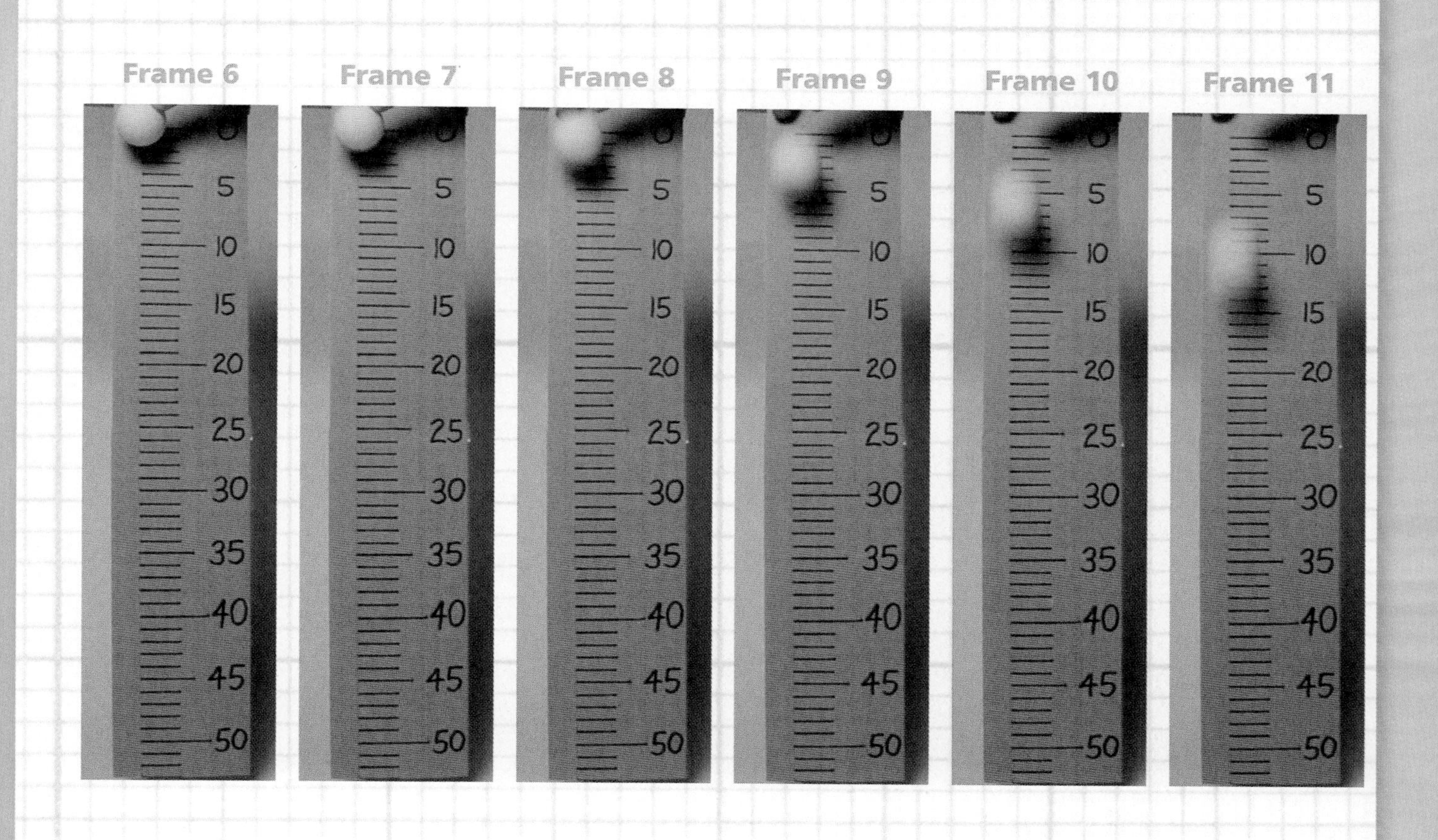

Frame 12

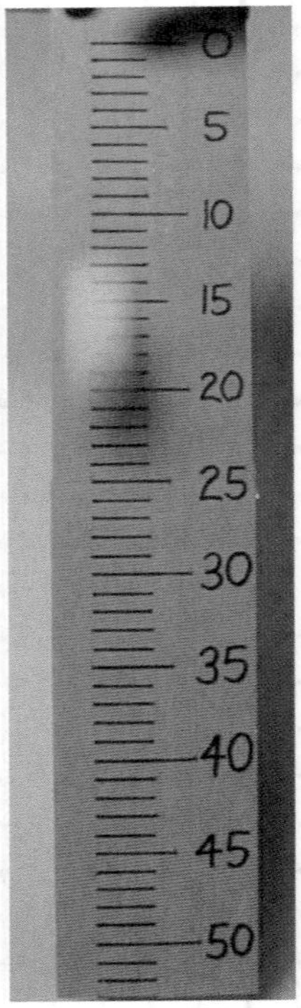

0 5 10 15 20 25 30 35 40 45 50

Frame 13

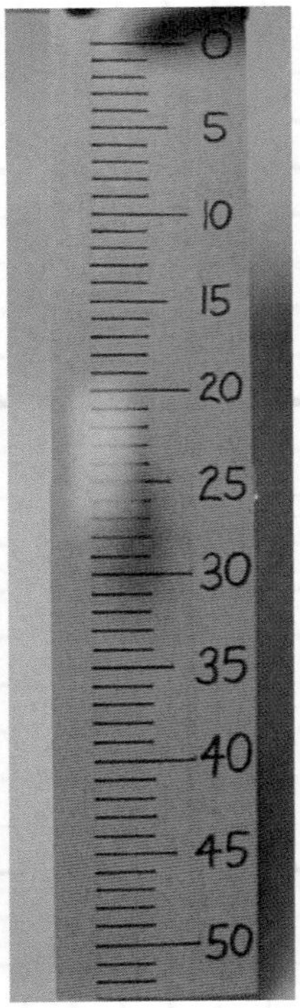

0 5 10 15 20 25 30 35 40 45 50

Frame 14

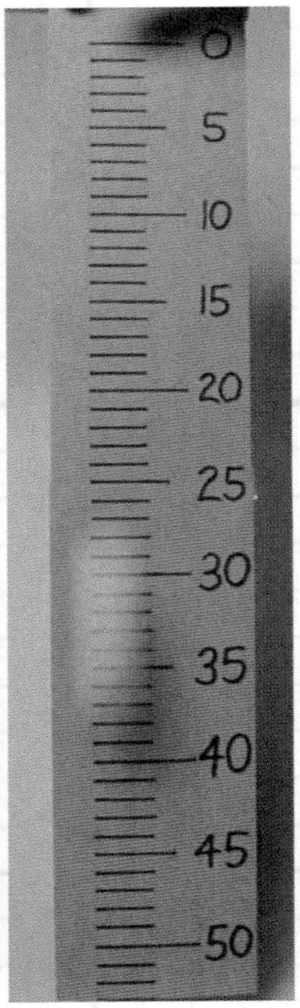

0 5 10 15 20 25 30 35 40 45 50

Frame 15

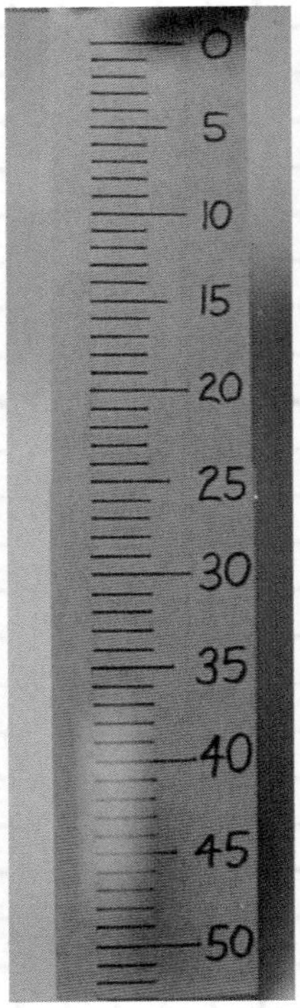

0 5 10 15 20 25 30 35 40 45 50

Frame 16

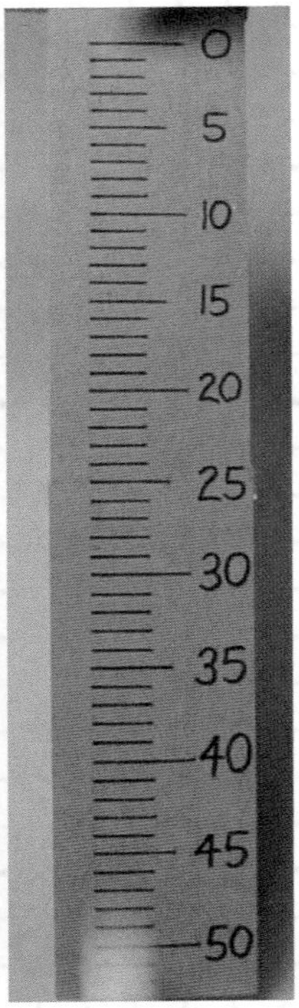

0 5 10 15 20 25 30 35 40 45 50

Acceleration of Gravity on Different Celestial Objects

Object	Classification	Surface Gravitational Acceleration (m/s^2)	Gravitational Force (Earth *g*)	Mass (Earth masses)	Radius (Earth radii)
Pluto	Dwarf planet	0.62	0.063	0.0022	0.18
Moon	Moon	1.6	0.16	0.012	0.27
Mercury	Planet	3.7	0.38	0.055	0.38
Mars	Planet	3.7	0.38	0.11	0.11
Venus	Planet	8.9	0.90	0.82	0.95
Earth	Planet	9.8	1	1	1
Uranus	Planet	8.7	0.89	15	4.0
Neptune	Planet	11.2	1.14	17	3.8
Saturn	Planet	10.4	1.07	95	9.1
Jupiter	Planet	24.8	2.5	318	11
Sun	Star	274.4	28	333,000	109

Challenge Questions

Why do you think that Mercury and Mars have the same gravitational acceleration, but different masses? What other information could you consider?

Why do you think that Uranus is more massive than Earth, but has less gravitational acceleration?

Why do you think that Saturn is more massive than Neptune, but has less gravitational acceleration?

Newton's Three Laws of Motion

In 1687, Sir Isaac Newton put forward his famous three laws of motion.

Law 1: Every object stays at rest unless an outside force acts on it. If an object is moving, it travels in a straight line at a constant speed unless an outside force acts on it.

Law 2: The acceleration of an object is directly proportional to the force on the object. It is inversely proportional to its mass ($a = F/m$, or $F = ma$).

Law 3: For every action or force, there is an equal and opposite reaction or force.

"Newton's Cradle" is a toy that demonstrates energy transfer. When you pull back and release the first ball, it strikes the second and stops. But its kinetic energy is transferred to the other balls. What happens next?

Engineering Design Process

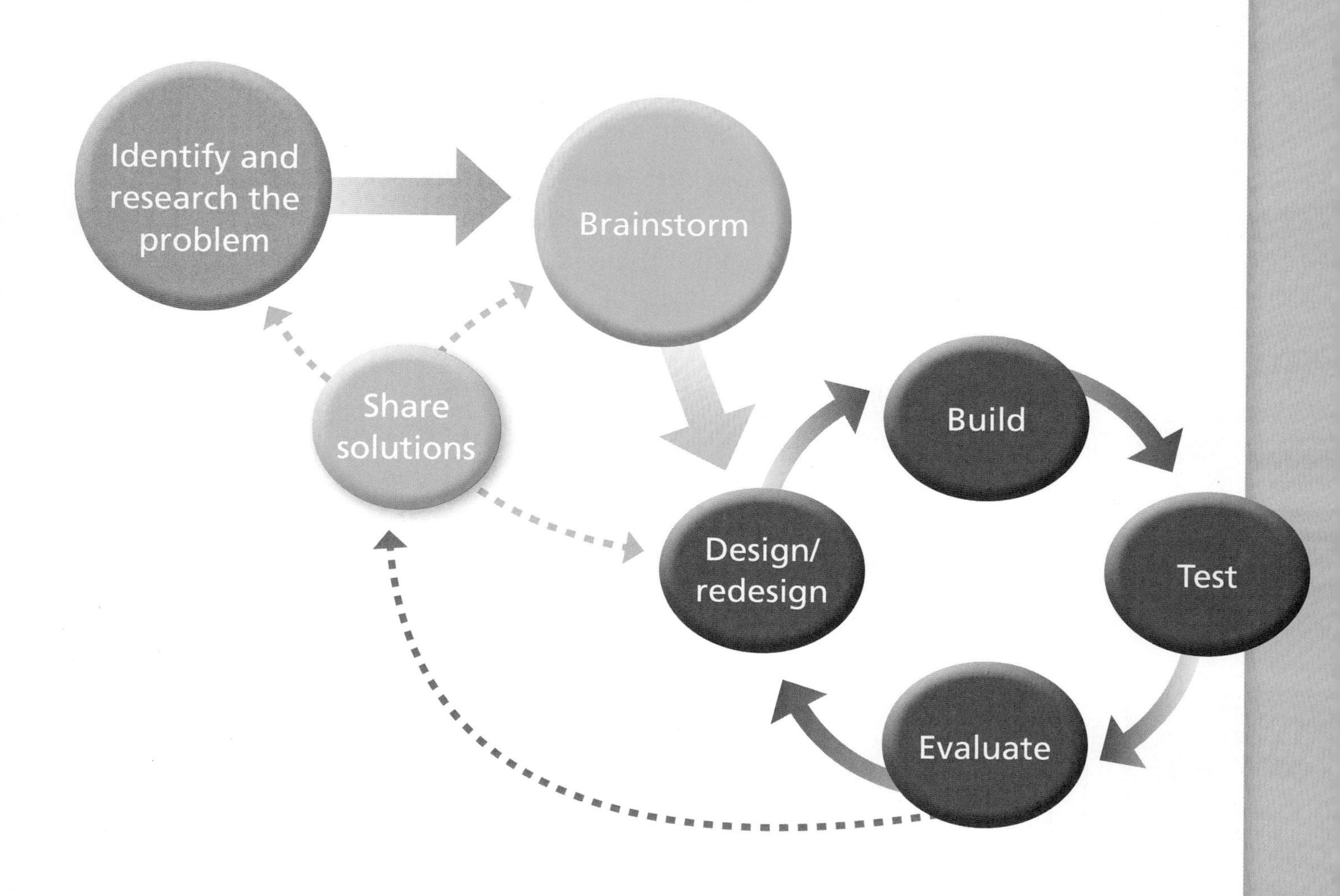

Science Practices

1. **Asking questions.** Scientists ask questions to guide their investigations. This helps them learn more about how the world works.
2. **Developing and using models.** Scientists develop models to represent how things work and to test their explanations.
3. **Planning and carrying out investigations.** Scientists plan and conduct investigations in the field and in laboratories. Their goal is to collect data that test their explanations.
4. **Analyzing and interpreting data.** Patterns and trends in data are not always obvious. Scientists make tables and graphs. They use statistical analysis to look for patterns.
5. **Using mathematics and computational thinking.** Scientists measure physical properties. They use computation and math to analyze data. They use mathematics to construct simulations, solve equations, and represent different variables.
6. **Constructing explanations.** Scientists construct explanations based on observations and data. An explanation becomes an accepted theory when there are many pieces of evidence to support it.
7. **Engaging in argument from evidence.** Scientists use argumentation to listen to, compare, and evaluate all possible explanations. Then they decide which best explains natural phenomena.
8. **Obtaining, evaluating, and communicating information.** Scientists must be able to communicate clearly. They must evaluate others' ideas. They must convince others to agree with their theories.

Are you a scientist?

Engineering Practices

1. **Defining problems.** Engineers ask questions to make sure they understand problems they are trying to solve. They need to understand the constraints that are placed on their designs.
2. **Developing and using models.** Engineers develop and use models to represent systems they are designing. Then they test their models before building the actual object or structure.
3. **Planning and carrying out investigations.** Engineers plan and conduct investigations. They need to make sure that their designed systems are durable, effective, and efficient.
4. **Analyzing and interpreting data.** Engineers collect and analyze data when they test their designs. They compare different solutions. They use the data to make sure that they match the given criteria and constraints.
5. **Using mathematics and computational thinking.** Engineers measure physical properties. They use computation and math to analyze data. They use mathematics to construct simulations, solve equations, and represent different variables.
6. **Designing solutions.** Engineers find solutions. They propose solutions based on desired function, cost, safety, how good it looks, and meeting legal requirements.
7. **Engaging in argument from evidence.** Engineers use argumentation to listen to, compare, and evaluate all possible ideas and methods to solve a problem.
8. **Obtaining, evaluating, and communicating information.** Engineers must be able to communicate clearly. They must evaluate other's ideas. They must convince others of the merits of their designs.

Are you an engineer?

Science Safety Rules

1. Always follow the safety procedures outlined by your teacher. Follow directions, and ask questions if you're unsure of what to do.
2. Never put any material in your mouth. Do not taste any material or chemical unless your teacher specifically tells you to do so.
3. Do not smell any unknown material. If your teacher asks you to smell a material, wave a hand over it to bring the scent toward your nose.
4. Avoid touching your face, mouth, ears, eyes, or nose while working with chemicals, plants, or animals. Tell your teacher if you have any allergies.
5. Always wash your hands with soap and warm water immediately after using chemicals (including common chemicals, such as salt and dyes) and handling natural materials or organisms.

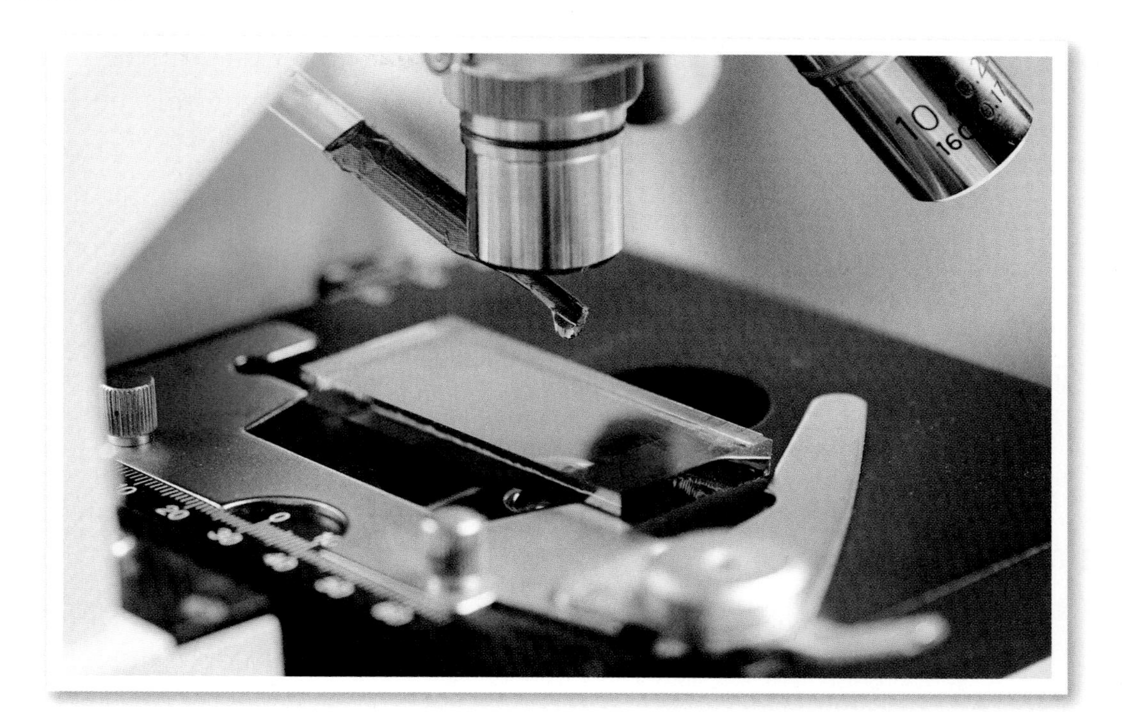

6. Do not mix unknown chemicals just to see what might happen.
7. Always wear safety goggles when working with liquids, chemicals, and sharp or pointed tools. Tell your teacher if you wear contact lenses.
8. Clean up spills immediately. Report all spills, accidents, and injuries to your teacher.
9. Treat animals with respect, caution, and consideration.
10. Never use the mirror of a microscope to reflect direct sunlight. The bright light can cause permanent eye damage.

Glossary

acceleration the change of velocity per unit of time

air resistance the force exerted by air molecules on objects moving through air

attraction to pull toward each other

average speed the theoretical constant speed at which an object would have to travel in order to go a given distance in a given period of time. Total distance divided by total time.

centripetal force a force acting on an object moving around another object, pulling it toward the center

change in position when an object moves from an initial position to a new position

collision when one object hits another object

concussion the most common form of brain injury

constant speed speed that does not vary over time

constraint a restriction or limitation

criterion (plural: criteria) a standard for evaluating or testing something

crumple zone a section of a vehicle designed to absorb energy during a collision and protect passengers

deceleration a negative change of velocity per unit of time (moving more slowly)

delta (Δ) a symbol that indicates change

distance how far between two objects or locations

energy usable power that can be transferred or converted to different forms, but cannot be created or destroyed

engineer a designer who uses scientific information and other considerations to accomplish a goal or solve a problem

escape velocity the minimum speed required to leave an orbit

field a space, or zone, around a mass

force an interaction between masses. A push or pull.

friction a force acting between surfaces in contact. Friction acts to resist motion.

gram (g) a unit of mass

gravitational force the force of attraction between objects with mass

gravity a force of attraction between masses

impulse force applied over a period of time

infer to reach a conclusion based on evidence

interval an amount, such as the time or distance between two markers. Standard interval units, like seconds and meters, are used to measure time and distance.

joule (j) a unit of kinetic energy

kinetic energy energy of motion

light speed the distance light can travel in a given time frame

mass a measure of the quantity of matter in an object

motion the process of changing position

net force the sum of all the forces acting on a mass

newton (N) a unit that describes measurement of force; the amount of force needed to accelerate 1 kg by 1 m/s^2

orbit the curved path of an object around another object

position the location of an object

potential energy energy stored in the position or condition of an object

rest not moving

slope in math or physics, the steepness of a line on a graph

speed the distance traveled by an object in a unit of time. Speed is reported in standard units of distance per unit time, such as meters per second or kilometers per hour.

speedometer a device that measures speed

supersonic speed when an object moves faster than the speed of sound

surface area a measurement of each face of an object's exterior dimensions

terminal velocity the maximum speed an object can obtain during free fall through air

theory of special relativity Albert Einstein's 1905 theory to explain the relationship between space and time

thermal energy radiant energy that heats

transfer to move from one to another

variable any factor that can be changed

weight a measurement of the force of gravity from a massive object (such as Earth) pulling another object toward it

The black marlin is the fastest recorded animal in water. They can grow up to 4.5 m long and weigh 454 kg, making them the most sought-after fish for sports fisherman to catch.

Index